D0659950

ELEMENTARY THEORY OF NUMBERS

by

WILLIAM J. LeVEQUE

University of Michigan

DOVER PUBLICATIONS, INC.

New York

Published in Canada by General Publishing Company, Ltd., 30 Lesmill Road, Don Mills, Toronto, Ontario.
Published in the United Kingdom by Constable and Company, Ltd.

This Dover edition, first published in 1990, is an unabridged, slightly corrected republication of the work originally published by Addison-Wesley Publishing Company, Inc., Reading, Mass., in 1962 in the "Addison-Wesley Series in Introductory Mathematics."
Manufactured in the United States of America
Dover Publications, Inc., 31 East 2nd Street, Mineola, N.Y. 11501

Library of Congress Cataloging-in-Publication Data

LeVeque, William Judson.
Elementary theory of numbers / William J. LeVeque.
p. cm.
Reprint. Originally published: Reading, Mass. : Addison-Wesley, 1962 (Addison-Wesley series in introductory mathematics)
Includes bibliographical references.
ISBN 0-486-66348-5
1. Number theory. I. Title. II. Series: Addison-Wesley series in introductory mathematics.
QA241.L57 1990
512'.7—dc20 90-2802
CIP

PREFACE

In the past few years there has been a great resurgence of interest in mathematics on both the secondary and undergraduate levels, and a growing recognition that the courses traditionally offered do not exhaust the mathematics which it is both possible and desirable to teach at those levels. Of course, not all of modern mathematics is accessible; some of it is too abstract to be comprehensible without more training in mathematical thinking, and some of it requires more technical knowledge than the young student can have mastered. Happily, the theory of numbers presents neither of these difficulties. The subject matter is the very concrete set of whole numbers, the rules are those the student has been accustomed to since grade school, and no assumption need be made as to special prior knowledge. To be sure, the results are not directly applicable in the physical world, but it is difficult to name a branch of mathematics in which the student encounters greater variety in types of proofs, or in which he will find more simple problems to stimulate his interest, challenge his ability, and increase his mathematical strength. For these and a number of other reasons, both the School Mathematics Study Group and the Committee on Undergraduate Programs have advocated the teaching of number theory to high-school and college students.

The present book is the result of an attempt to expose the subject in such form as to be accessible to persons with less mathematical training than those who would normally read, for example, the author's *Topics in Number Theory*, Volume I. There is considerable overlapping in material, of course—it is, after all, the same subject—but the exposition is more leisurely, the examples and computational problems are more numerous, and certain relatively difficult topics have been omitted. Furthermore, the chapter on Gaussian arithmetic is entirely new, and the chapters on continued fractions and Diophantine equations have been almost entirely rewritten. I hope that the book may prove useful in high-school enrichment programs, in nontraditional freshman and sophomore courses, and in teacher training and refresher programs.

Certain problems are starred to indicate greater than average difficulty.

W. J. L.

CONTENTS

CHAPTER 1

INTRODUCTION

1–1 What is number theory? In number theory we are concerned with properties of certain of the integers (whole numbers)

$$\ldots, -3, -2, -1, 0, 1, 2, 3, \ldots,$$

or sometimes with those properties of real or complex numbers which depend rather directly on the integers. It might be thought that there is little more that can be said about such simple mathematical objects than what has already been said in elementary arithmetic, but if you stop to think for a moment, you will realize that heretofore integers have not been considered as interesting objects in their own right, but simply as useful carriers of information. After totaling a grocery bill, you are interested in the amount of money involved, and not in the number representing that amount of money. In considering sin 31°, you think either of an angular opening of a certain size, and the ratios of some lengths related to that angle, or of a certain position in a table of trigonometric functions, but not of any interesting properties that the number 31 might possess.

The attitude which will govern the treatment of integers in this text is perhaps best exemplified by a story told by G. H. Hardy, an eminent British number theorist who died in 1947. Hardy had a young protégé, an Indian named Srinivasa Ramanujan, who had such a truly remarkable insight into hidden arithmetical relationships that, although he was almost uneducated mathematically, he did a great amount of first-rate original research in mathematics. Ramanujan was ill in a hospital in England, and Hardy went to visit him. When he arrived, he idly remarked that the taxi in which he had ridden had the license number 1729, which, he said, seemed to him a rather uninteresting number. Ramanujan immediately replied that, on the contrary, 1729 was singularly interesting, being the smallest positive integer expressible as a sum of two positive cubes in two different ways, namely $1729 = 10^3 + 9^3 = 12^3 + 1^3$!

It should not be inferred that one needs to know all such little facts to understand number theory, or that one needs to be a lightning calculator; we simply wished to make the point that the question of what the smallest integer is which can be represented as a sum of cubes in two ways is of interest to a number theorist. It is interesting not so much for its own sake (after all, anyone could find the answer after a few minutes of unimaginative computation), but because it raises all sorts of further questions whose answers are by no means simple matters of calculation. For example, if s is any positive integer, about how large is the smallest integer representable as a sum of cubes of positive integers in s different ways? Or, are there infinitely many integers representable as a sum of cubes in two different ways? Or, how can one characterize in a different fashion the integers which can be represented as a sum of two cubes in at least one way? Or, are any *cubes* representable as a sum of two cubes? That is, has the equation

$$x^3 + y^3 = z^3 \tag{1}$$

any solutions in positive integers x, y, and z? These questions, like that discussed by Hardy and Ramanujan, are concerned with integers, but they also have an additional element which somehow makes them more significant: they are concerned not with a particular integer, but with whole classes or collections of integers. It is this feature of generality, perhaps, which distinguishes the theory of numbers from simple arithmetic. Still, there is a gradual shading from one into the other, and number theory is, appropriately enough, sometimes called higher arithmetic.

In view of the apparent simplicity of the subject matter, it is not surprising that number-theoretic questions have been considered throughout almost the entire history of recorded mathematics. One of the earliest such problems must have been that of solving the "Pythagorean" equation

$$x^2 + y^2 = z^2. \tag{2}$$

For centuries it was supposed that the classical theorem embodied in (2) concerning the sides of a right triangle was due either to Pythagoras or a member of his school (about 550 B.C). Recently interpreted cuneiform texts give strong evidence, however, that Babylonian mathematicians not only knew the theorem as early as 1600 B.C., but that they knew how to compute all integral solutions x, y, z of (2), and used this knowledge for the construction of crude trigonometric tables. There is no difficulty in finding a large number of integral solutions of (2) by trial and error—just add many different pairs of squares, and some of the sums will turn out to be squares also. Finding *all* solutions is another matter, requiring understanding rather than patience. We shall treat this question in detail in Chapter 5.

Whatever the Babylonians may have known and understood, it seems clear that we are indebted to the Greeks for their conception of mathematics as a systematic theory founded on axioms or unproved assumptions, developed by logical deduction and supported by strict proofs. It would probably not have occurred to the Babylonians to write out a detailed analysis of the integral solutions of (2), as Euclid did in the tenth book of his *Elements*. This contribution by Euclid was minor, however, compared with his invention of what is now called the Euclidean algorithm, which we shall consider in the next chapter. Almost equally interesting was his proof that there are infinitely many *prime numbers*, a prime number being an integer such as 2, 3, 5, etc., which has no exact divisors except itself, 1, and the negatives of these two numbers.* We shall repeat this proof later in the present chapter.

Another Greek mathematician whose work remains significant in present-day number theory is Diophantos, who lived in Alexandria, about 250 A.D. Many of his writings have been lost, but they all seem to have been concerned with the solution in integers (or sometimes in rational numbers) of various algebraic equations. In his honor we still refer to such equations as (1) and (2) above as Diophantine equations, not because they are special kinds of equations, but because special kinds of solutions are required. Diophantos considered a large number of such equations, and his work was continued by the Arabian Al-Karkhi (ca. 1030) and the Italian Leonardo Pisano (ca. 1200). Although it is possible that these latter works were known to Pierre Fermat (1601–1665), the founding father of number theory as a systematic branch of knowledge, it is certain that Fermat's principal inspiration came directly from Diophantos' works.

The questions considered in the theory of numbers can be grouped according to a more or less rough classification, as will now be explained. It should not be inferred that every problem falls neatly into one of these classes, but simply that many questions of each of the following categories have been considered.

First, there are multiplicative problems, concerned with the divisibility properties of integers. It will be proved later that any positive integer n greater than 1 can be represented uniquely, except for the order of the factors, as a product of one or more positive primes. For example,

$$12 = 2 \cdot 2 \cdot 3, \qquad 13 = 13, \qquad 2{,}892{,}384 = 2^5 \cdot 3^2 \cdot 11^2 \cdot 83,$$

and there is no essentially different factorization of these integers, if the factors are required to be primes. This unique factorization theorem, as it is called, might almost be termed the fundamental theorem of number theory, so manifold and varied are its applications. From the decomposition of n into primes, it is easy to determine the number of positive divisors (i.e., exact divisors) of n. This number, which of course depends on n, is called $\tau(n)$ by some writers and $d(n)$ by others; we shall use the former designation (τ is the Greek letter *tau;* see the Greek alphabet in the ap-

---

* The term *prime* will usually be reserved for the positive integers with this property; the numbers −2, −3, −5, etc., will be called *negative primes.* Note that 1 is not included among the primes.

TABLE 1–1

n	$\tau(n)$	n	$\tau(n)$
1	1	13	2
2	2	14	4
3	2	15	4
4	3	16	5
5	2	17	2
6	4	18	6
7	2	19	2
8	4	20	6
9	3	21	4
10	4	22	4
11	2	23	2
12	6	24	8

pendix). The behavior of $\tau(n)$ is very erratic, as we can see by examining Table 1–1. If $n = 2^m$, the divisors of n are $1, 2, 2^2, \ldots, 2^m$, so that $\tau(2^m) = m + 1$. On the other hand, if n is a prime, then $\tau(n) = 2$. Since, as we shall see, there are infinitely many primes, it appears that the τ-function has arbitrarily large values, and yet has the value 2 for infinitely many n. A number of questions might occur to anyone who thinks about the subject for a few moments and studies the above table. For example:

(a) Is it true that $\tau(n)$ is odd if and only if n is a square?

(b) Is it always true that if m and n have no common factor larger than 1, then $\tau(m)\tau(n) = \tau(mn)$?

(c) How large can $\tau(n)$ be in comparison with n? From the equation

$$\tau(2^m) = m + 1 = \frac{\log 2^m}{\log 2} + 1,$$

it might be guessed that perhaps there is a constant c such that

$$\tau(n) < c \log n \tag{3}$$

for all n. If this is false, is there any better upper bound than the trivial one, $\tau(n) \leq n$? (The last inequality is a consequence of the fact that only the n integers $1, 2, \ldots, n$ could possibly divide n.)

(d) How large is $\tau(n)$ on the average? That is, what can be said about the quantity

$$\frac{1}{N}\left(\tau(1) + \tau(2) + \cdots + \tau(N)\right) \tag{4}$$

as N increases indefinitely?

(e) For large N, approximately how many solutions $n \leq N$ are there of the equation $\tau(n) = 2$? In other words, about how many primes are there among the integers $1, 2, \ldots, N$?

Of these questions, which are fairly typical problems in multiplicative number theory, the first two are very easy to answer in the affirmative. The third and fourth are somewhat more difficult, and we shall not consider them further in this book. However, to satisfy the reader's curiosity, we mention that (3) is false for certain sufficiently large n, however large the constant c may be, whereas the inequality $\tau(n) < cn^{\epsilon}$ is true for all sufficiently large n, however small the positive constants c and ϵ may be. On the other hand, for large N, the average value (4) is nearly equal* to $\log N$. The last is very difficult indeed. It was conjectured independently by C. F. Gauss and A. Legendre, in about 1800, that the number, commonly called $\pi(N)$, of primes not exceeding N is approximately $N/\log N$, in the sense that the relative error

$$\frac{|\pi(N) - (N/\log N)|}{N/\log N} = \left|\frac{\pi(N)}{N/\log N} - 1\right|$$

is very small when N is sufficiently large. Many years later (1852–54), P. L. Chebyshev showed that if this relative error has any limiting value, it must be zero, but it was not until 1896 that J. Hadamard and C. de la Vallée Poussin finally proved what is now called the prime number theorem, that

$$\lim_{N\to\infty} \frac{\pi(N)}{N/\log N} = 1.$$

In 1948 a much more elementary proof of this theorem was discovered by the Norwegian mathematician Atle Selberg and the Hungarian mathematician Paul Erdös; even this proof, however, is too difficult for inclusion here.

Secondly, there are the problems of additive number theory: questions concerning the representability, and number of representations, of a positive integer as a sum of integers of a specified kind. For example, certain integers such as $5 = 1^2 + 2^2$ and $13 = 2^2 + 3^2$ are representable as a sum of two squares, and some, such as $65 = 1^2 + 8^2 = 4^2 + 7^2$, have two such representations, while others, such as 6, have none. Which integers are so representable, and how many representations are there? If we use four squares instead of two, we obtain Table 1–2.

* In the sense that the relative error is small. Here and elsewhere the logarithm is the so-called natural logarithm, which is a certain constant 2.303... times the logarithm to the base ten.

TABLE 1-2

$1 = 1^2 + 0^2 + 0^2 + 0^2$	$11 = 3^2 + 1^2 + 1^2 + 0^2$
$2 = 1^2 + 1^2 + 0^2 + 0^2$	$12 = 2^2 + 2^2 + 2^2 + 0^2$
$3 = 1^2 + 1^2 + 1^2 + 0^2$	$13 = 3^2 + 2^2 + 0^2 + 0^2$
$4 = 2^2 + 0^2 + 0^2 + 0^2$	$14 = 3^2 + 2^2 + 1^2 + 0^2$
$5 = 2^2 + 1^2 + 0^2 + 0^2$	$15 = 3^2 + 2^2 + 1^2 + 1^2$
$6 = 2^2 + 1^2 + 1^2 + 0^2$	$16 = 4^2 + 0^2 + 0^2 + 0^2$
$7 = 2^2 + 1^2 + 1^2 + 1^2$	$17 = 4^2 + 1^2 + 0^2 + 0^2$
$8 = 2^2 + 2^2 + 0^2 + 0^2$	$18 = 3^2 + 3^2 + 0^2 + 0^2$
$9 = 3^2 + 0^2 + 0^2 + 0^2$	$19 = 3^2 + 3^2 + 1^2 + 0^2$
$10 = 3^2 + 1^2 + 0^2 + 0^2$	$20 = 4^2 + 2^2 + 0^2 + 0^2$

From this or a more extensive table, it is reasonable to guess that every positive integer is representable as a sum of four squares of nonnegative integers. This is indeed a correct guess, which seems already to have been made by Diophantos. A proof was known to Pierre Fermat in 1636, but the first published proof was given by Joseph Louis Lagrange in 1770.

More generally, it was proved by David Hilbert in 1909 that if we consider kth powers rather than squares, a certain fixed number of them suffices for the representation of any positive integer.

There are also some very interesting questions about sums of primes. It was conjectured by Charles Goldbach in 1742 that every even integer larger than 4 is the sum of two odd primes. (All primes except 2 are odd, of course, since evenness means divisibility by 2.) Despite enormous efforts in the 200 intervening years by many excellent mathematicians, the truth or falsity of Goldbach's conjecture has not been settled to this day. It is known, however, that every odd integer larger than $10^{350,000}$ is the sum of three odd primes, which implies that every even integer larger than this same number is the sum of four primes. It has also been conjectured, so far without proof, that every even integer is representable in infinitely many ways as the difference of two primes. In particular, this would mean that there are infinitely many *prime twins*, such as 17 and 19, or 101 and 103, which differ by 2.

As a third class of problems, there are the Diophantine equations mentioned earlier. Here the general theory is rather scanty, since the subject is intrinsically very difficult. In Chapter 3 we shall give a complete analysis of the linear equation in two unknowns, $ax + by = c$; that is, we shall determine the exact conditions which a, b, and c must satisfy in order that the equation be solvable in integers, and we shall present an effective procedure for finding these solutions. Certain quadratic equations, such as the Pythagorean equation $x^2 + y^2 = z^2$ and the so-called Pell equation, $x^2 - dy^2 = 1$, can also be solved completely, but relatively little is known about higher-degree equations in general, although certain specific equations have been solved. For example, there is the conjecture due to Fermat, that the equation $x^n + y^n = z^n$ has no solutions in non-zero integers x, y, and z if $n > 2$. This is perhaps the oldest and best-known unsolved problem in mathematics. The conjecture is now known to be correct for all $n < 4000$, and it is also known that if n is a prime smaller than 253,747,889, then there is no solution in which none of x, y, or z is divisible by n; but the general proposition remains out of reach.

There are many other branches of number theory—Diophantine approximation, geometry of numbers, theory of quadratic forms, and analytic theory of numbers, to name a few—but their descriptions are more complicated, and since we shall not consider problems from such fields in this book, we shall not enter into details. In any case, no classification can be exhaustive, and perhaps enough examples have been given to show the typical flavor of number-theoretic questions.

Granting that the reader now knows what number theory is, or that he will after reading this book, there is still the question of why anyone should create or study the subject. Certainly not because of its applicability to problems concerning the physical world; such applications are extremely rare. The theory of numbers has, on the other hand, been a strong influence in the development of higher pure mathematics, both in stimulating the creation of powerful general methods in the course of solving special problems (such as the Fermat conjecture above, and the prime number theorem) and as a source of ideas and inspiration comparable to geometry and the mathematics of physical phenomena; and so in retrospect it turns out to have been worth developing. But these were not the reasons that led men to ponder arithmetical questions, in former times, nor are they the reasons for the present day interest in the theory of numbers. The driving force is rather man's insatiable curiosity—the drive to know and do everything. In the case at hand this curiosity is whetted considerably by the surprising difficulty of the subject, maintained by its tremendous diversity, and rewarded by the elegance and unexpectedness of the results. It is these attributes, perhaps, which led Carl Friedrich Gauss (1777–1855), one of the two or three greatest mathematicians who ever lived, to label the theory of numbers the Queen of mathematics.

1–2 Foundations. In the remaining chapters of this book we shall adopt the attitude that the integers and the basic arithmetical operations by means of which they are combined have already been comprehended by the reader, and we shall not dwell on such questions as what the integers are, or why $2 + 2 = 4$ and $2 + 3 = 3 + 2$. Detailed logical developments of these topics exist (see, for example, E. Landau's *Founda-*

tions of Analysis, Chelsea Publ. Co., New York, 1951), and anyone seriously interested in mathematics should examine a book on this subject at some time, to see what is really behind the arithmetic and elementary algebra he learned in grade school. That is not our objective here, however, and in this section and the next two we shall simply single out a few matters which may be genuinely new to the student, and review the remainder very quickly.

The arithmetic of the integers, like the geometry of the plane, can be made to depend on a few axioms, in the sense that everything else follows from them by accepted logical rules. One such set of axioms was given by G. Peano in 1889; it characterizes the set (class, collection) of natural numbers 1, 2, 3, . . . , and consists of the following postulates:

(1) 1 is a natural number.
(2) To each natural number x there corresponds a second natural number x', called the successor of x.
(3) 1 is not the successor of any natural number.
(4) From $x' = y'$ follows $x = y$.
(5) Let M be a set of natural numbers with the following two properties:
   (a) 1 belongs to M.
   (b) If x belongs to M, then x' also belongs to M.

   Then M contains all natural numbers.

In the language of these axioms, addition is defined by setting $x + 1 = x'$, $x + 2 = (x')'$, etc., and multiplication is defined in terms of addition: $ab = a + a + \cdots + a$, where there are b terms on the right. The usual rules of algebra can then be deduced, as they apply to the natural numbers, and the inequality symbol "$<$" can be introduced. Finally, zero and the negative integers are defined in terms of the natural numbers by various devices.

This path is rather long, and it might be worth while also to list a second set of axioms, which become theorems if one starts from the Peano postulates, and which are more numerous and complicated than the latter, but which relate more directly to the workaday world of algebra and arithmetic and, in addition, suffice to deduce all further rules in these subjects.

I. Each pair of integers a and b has a unique sum $a + b$ and a unique product $a \cdot b$ or ab, such that the following associative, commutative, and distributive laws hold:

	Addition	*Multiplication*
Associative law:	$a + (b + c) = (a + b) + c$	$a(bc) = (ab)c$
Commutative law:	$a + b = b + a$	$ab = ba$
Distributive law:	$a(b + c) = ab + ac$	

II. The distinct integers 0 and 1 have the properties that $a + 0 = a$ and $1 \cdot a = a$ for all integers a.
III. For each integer a the equation $a + x = 0$ has a unique solution x, which we call $-a$.
IV. If $c \neq 0$ and $ca = cb$, then $a = b$.
V. There is a subset of integers, called positive integers, with the following properties: the sum and product of positive integers are positive, and every nonzero integer a has the property that exactly one of the two numbers a and $-a$ is positive.

The notion of inequality can be introduced in terms of the positive integers: we say that a is smaller than b, and write $a < b$, if $b - a$ is a positive integer, and we write $a \leq b$ if either $a < b$ or $a = b$. It can then be proved that if $a < b$, then $a + c < b + c$ for all integers c, and if $a < b$ and $0 < c$, then $ac < bc$. The absolute value $|a|$ of an integer a is 0 if $a = 0$; otherwise it is the positive one of the two numbers a, $-a$.

VI. Every set of positive integers which contains at least one member contains a smallest member. That is, there is an integer a in the set such that $a \leq b$ for every b in the set.

Presumably the first five axioms of this second set are already familiar to the reader as working rules, although perhaps not in the explicit form presented here. The sixth could hardly be surprising, as a "fact" about the integers, but it may be surprising how useful it is as a device for proving theorems. We shall devote the next section to that topic, but we give a first example of it now by showing how the last Peano postulate follows from the above six axioms.

First we prove that *there is no integer between* 0 *and* 1. For if there is at least one such integer, then the set, call it A, of integers a such that $0 < a < 1$, has at least one member. (More briefly, we say that A is not empty.) By Axiom VI, A has a smallest element, say b. But by multiplying through in the inequality $0 < b < 1$ by the positive number b we obtain $0 < b^2 < b$, so that b^2 is also an element of A, and is smaller than b, contrary to the definition of b. Hence A cannot contain a smallest element, and so must be empty. ▲*

We can now deduce the fifth Peano postulate from Axiom VI above. Suppose that M is a set of integers having the two properties described in Postulate 5, and let S be the set of all positive integers not in M—the so-called *complementary* set to M. It suffices to show that S is empty, to see that M contains all positive integers. Suppose that S contains at least one element; then it contains a smallest element, say d. But $d \neq 1$, by the first property in Postulate 5, and since there is no positive integer

* This symbol signals the end of a proof.

between 0 and 1, we have $d > 1$. But then $d - 1$ is positive, and since d is the smallest positive integer in S, the smaller number $d - 1$ must be in M. But by the second property in Postulate 5, $(d - 1) + 1 = d$ must also be in M, which is false. Having arrived at a contradiction, our assumptions must have been inconsistent with one another, so that if M is in fact a set having the properties listed in Postulate 5, then its complementary set must be empty, and M must contain all positive integers. ▲

It is customary to call Postulate 5 the *induction axiom*, and Axiom VI the *well-ordering axiom*. (A set of numbers is said to be well ordered if every subset has a smallest element.) We have just seen that the second implies the first (if Axioms I through V are assumed), and conversely, it is possible to show that the first implies the second. They are therefore different versions of the same principle, and can be used interchangeably, as we shall now see.

1–3 Proofs by induction. On many occasions, both in this book and in the student's later mathematical work, a theorem must be proved which is of the form, "For every positive integer n, the sentence $P(n)$ is true." Here we have used $P(n)$ as a name for some sentence or other which involves an integer-valued variable n. For example, $P(n)$ might be "n is the sum of the squares of four nonnegative integers," or "the sum of the integers from 1 to n inclusive is $n(n + 1)/2$," or "$(1 + x)^n > 1 + nx$ for $x > -1$." The induction axiom can frequently be used to prove such theorems in the following way. In the fifth Peano postulate we take for M the set of positive integers n for which $P(n)$ is true; then showing that M contains all positive integers amounts to showing that $P(n)$ is true for all n, as asserted. What must be done, then, is to show that M contains 1 [i.e., that $P(1)$ is a true sentence] and that M contains the successor of each of its elements [i.e., that if $P(m)$ is true, so also is $P(m + 1)$, for every positive integer m.]

Let us apply this to the proof of the second example above:

For every positive integer n, we have

$$1 + 2 + 3 + \cdots + n = \frac{n(n+1)}{2}.$$

We must first show that $P(1)$ is true, which is obvious:

$$1 = \frac{1(1+1)}{2}.$$

Next, suppose that m is an integer such that

$$1 + 2 + \cdots + m = \frac{m(m+1)}{2}.$$

If we add $m + 1$ to both sides of this true equation, we obtain

$$1 + 2 + \cdots + m + (m + 1) = \frac{m(m + 1)}{2} + (m + 1)$$
$$= (m + 1)\left(\frac{m}{2} + 1\right),$$

so that

$$1 + 2 + \cdots + m + (m + 1) = \frac{(m + 1)(m + 2)}{2},$$

which is exactly the sentence $P(m + 1)$. Thus $P(m + 1)$ is true whenever $P(m)$ is, and hence by the induction axiom, $P(n)$ is true for all positive integers n. ▲

As a second example, consider the Fibonacci sequence

$$1, 1, 2, 3, 5, 8, 13, 21, \ldots,$$

in which every element after the second is the sum of the two numbers immediately preceding it. (In general, a *sequence* is an ordered array of elements, having a first element, a second element, etc.) If we denote by u_n the nth element of this sequence, then the sequence is defined by the conditions

$$u_1 = 1,$$
$$u_2 = 1,$$
$$u_n = u_{n-1} + u_{n-2}, \qquad n \geq 3. \tag{5}$$

(This is an example of a *recursive* definition, in which infinitely many elements are defined, each element, after a certain point, being defined in terms of preceding elements.) Consider the following theorem: no two successive numbers u_n and u_{n+1} have a common factor greater than 1. This can be rephrased in the form considered above, as follows:

For every positive integer n, u_n and u_{n+1} have no common factor greater than 1.

Clearly $P(1)$ is true: 1 and 1 have no common factor greater than 1. Suppose now that m is any integer for which $P(m)$ is true: u_m and u_{m+1} have no common factor. (Henceforth the restriction "greater than 1" will be understood.) Then $P(m + 1)$ cannot be false, for if it were, u_{m+1} and u_{m+2} would have a common factor, say d, and this would also be a factor of u_m, since

$$u_m = u_{m+2} - u_{m+1}.$$

But then d would be a common factor of u_m and u_{m+1}, which is not the case. Hence the truth of $P(m)$ implies that of $P(m + 1)$ for every positive integer m, and the induction axiom shows that the theorem displayed above is true. ▲

This proof can easily be recast in terms of the well-ordering axiom. If the theorem is false, there is an n for which u_n and u_{n+1} have a common factor, and hence a smallest such n, say $n = m$. Now m is not 1, since u_1 and u_2 have no common factor, and therefore $m > 1$. But if u_m and u_{m+1} have the common factor d, then, by our previous reasoning, d is also a factor of u_{m-1}, and so u_m and u_{m-1} have a common factor. But this contradicts the definition of m. Hence the theorem is not false. ▲

There are two useful variations on the induction principle which should be noted. In one, the theorem to be proved concerns the integers not less than some fixed integer n_0, rather than those not less than 1, and it is easily seen that the induction axiom implies the following principle:

> If $P(n)$ is true for n_0, and if $P(m + 1)$ is true for every integer $m \geq n_0$ for which $P(m)$ is true, then $P(n)$ is true for every integer $n \geq n_0$.

The second variation allows for the case in which $P(m + 1)$ cannot easily be deduced from $P(m)$, but depends instead on $P(k)$ for some $k < m$:

> If $P(1)$ is true, and if $P(m + 1)$ is true for every $m \geq 1$ for which all of $P(1), P(2), \ldots, P(m)$ are true, then $P(n)$ is true for every $n \geq 1$.

This version is obtained from Postulate 5 by taking for M the set of positive integers for which all of $P(1), P(2), \ldots, P(n)$ are true.

To illustrate the use of the last version, consider the following theorem:

> *For every positive integer n, $u_n < (\frac{7}{4})^n$.*

Let $P(n)$ be the inequality (it is a sentence) "$u_n < (\frac{7}{4})^n$", and for brevity set $\alpha = \frac{7}{4}$. Then $P(1)$ and $P(2)$ are certainly true: $u_1 < \alpha$ and $u_2 < \alpha^2$. But from the truth of $P(m)$ we cannot deduce that of $P(m + 1)$ directly, since from $u_m < \alpha^m$ it follows only that $u_{m-1} < \alpha^m$ also, and hence that

$$u_{m+1} = u_m + u_{m-1} < \alpha^m + \alpha^m = 2\alpha^m = \tfrac{8}{7}\,\alpha^{m+1},$$

which is not the inequality we need. On the other hand, if we suppose that $P(1), \ldots, P(m)$ are all true, then we have for $m \geq 2$,

$$u_{m+1} = u_m + u_{m-1} < \alpha^m + \alpha^{m-1} = \alpha^{m-1}(1 + \alpha) < \alpha^{m-1} \cdot \alpha^2 = \alpha^{m+1},$$

since $1 + \alpha = \frac{11}{4} < 3 < \frac{49}{16} = \alpha^2$. ▲

There are then two parts to a proof by induction: verification of the sentence $P(n)$ for the smallest relevant value n_0 of the so called "induction variable" n, and proof of a certain implication, either "$P(m)$ implies $P(m + 1)$ for all integers $m \geq n_0$" or "[$P(1)$ and $P(2)$ and ... and $P(m)$]

implies $(m+1)$ for all integers $m \geq n_0$." Each of these implications has, in common with all implications, an hypothesis and a conclusion, and in the present context the former is called the "induction hypothesis." The student should note carefully the difference between the theorem to be proved, "$P(n)$ is true for all positive integers n," and the induction step, "$P(m)$ implies $P(m+1)$ for all positive integers m." They are both assertions about a sentence $P(n)$, but they are not the same. Moreover, although we frequently assume that $P(n)$ is true when we prove the induction step, this is not the same thing as assuming the truth of the theorem (which would of course make the whole thing nonsense), since the theorem is not the sentence $P(n)$ but the next to last sentence in quotation marks above about $P(n)$. Strictly speaking, $P(n)$ by itself is neither true nor false (because it contains the variable n), and does not become so until n is assigned a value. (The inequality "$2 > 3$" is false, the inequality "$5 > 3$" is true; the inequality "$n > 3$" is neither.) When we say, "Suppose that $P(m)$ is true," we really mean, "Suppose that m is an integer such that $P(m)$ is true."

Problems

When a number of quantities of the same general form are to be added or multiplied together, it is customary to abbreviate the sum and the product in the following way:

$$a_1 + a_2 + \cdots + a_n = \sum_{k=1}^{n} a_k,$$

$$a_1 a_2 \cdots a_n = \prod_{k=1}^{n} a_k.$$

Thus we would write $\tau(1) + \tau(2) + \cdots + \tau(n)$ as

$$\sum_{k=1}^{n} \tau(k),$$

and could define the factorial $n!$ by the equations

$$n! = \begin{cases} 1 & \text{for} \quad n = 0, \\ \prod_{k=1}^{n} k & \text{for} \quad n \text{ a positive integer.} \end{cases}$$

1. Prove the following identities by induction:

(a) $\displaystyle\sum_{k=1}^{n} k^2 = \frac{n(n+1)(2n+1)}{6}$; (b) $\displaystyle\sum_{k=1}^{n} k^3 = \frac{n^2(n+1)^2}{4}$;

(c) $\prod_{k=1}^{n}\left(1+\frac{1}{k}\right) = n+1;$ (d) $\sum_{k=1}^{n}(2k-1) = n^2.$

2. Show that if n is a positive integer and x is a real number larger than -1, then $(1+x)^n > 1+nx$.

3. Show that every integer greater than 1 can be represented as a product of one or more primes.

4. Show that if a and b are positive integers, there is a positive integer n such that $na > b$. [*Hint:* consider the differences $b - na$, and apply the well-ordering axiom.]

5. Define the *binomial coefficients* $\binom{n}{k}$ for integers n and k with $0 \le k \le n$ by the equation

$$\binom{n}{k} = \frac{n!}{k!(n-k)!}.$$

Show by direct computation that

$$\binom{n}{k} + \binom{n}{k+1} = \binom{n+1}{k+1} \quad \text{for} \quad 0 \le k < n.$$

Use this identity to prove the *binomial theorem* by induction:

$$(a+b)^n = a^n + \binom{n}{1}a^{n-1}b + \binom{n}{2}a^{n-2}b^2 + \cdots + \binom{n}{n-1}ab^{n-1} + b^n$$

$$= \sum_{k=0}^{n}\binom{n}{k}a^{n-k}b^k \quad \text{for positive integer } n \text{ and arbitrary } a \text{ and } b.$$

6. Show that a set S of n distinguishable elements has exactly 2^n subsets, including the empty set and S itself.

The assertion is sometimes made that mathematical induction is not useful for discovering new information, but only for verifying what has already been guessed. The following problems bear on this point.

7. Re-examine the proof that $u_n < \alpha^n$ for all n, to find the smallest β such that you could prove that $u_n < \beta^n$ for all n, and carry out this proof. Can you prove an inequality in the opposite direction, of the form $u_n > c\beta^n$, for some positive constant c?

8. From the binomial theorem we have

$$(1+1)^{2n} = \binom{2n}{0} + \binom{2n}{1} + \cdots + \binom{2n}{n} + \cdots + \binom{2n}{2n},$$

and hence $\binom{2n}{n} < 2^{2n} = 4^n$. Using the definition of the binomial coefficients, show that

$$\binom{2n+2}{n+1} = 4\left(1 - \frac{1}{2n+2}\right)\binom{2n}{n}.$$

Deduce an inequality of the form

$$(*) \qquad \binom{2n}{n} \geq c\beta^n$$

which is valid for $n \geq 1$, where c and β are specific numbers. Show also that for every $\beta < 4$ there is a positive constant c such that (*) holds for $n \geq 1$.

1-4 Indirect proofs. There is a second kind of proof with which the reader may not have had much experience, the so-called *indirect* proof, or proof *by contradiction*. An assertion P is said to have been proved by contradiction if it has been shown that, by assuming P to be false, we can deduce an assertion Q which is known to be false or which contradicts the assumption that P is false. Several proofs by contradiction have already occurred above; for example, the proof that there is no integer between 0 and 1 and the deduction of the induction axiom from the well-ordering axiom were indirect. As another example, consider the theorem (known as early as the time of Euclid) that there are infinitely many prime numbers. To prove this by contradiction, we assume the opposite, namely that there are only finitely many prime numbers. Let these be $p_1, \ldots, p_n$; let N be the integer $p_1p_2 \cdots p_n + 1$; and let Q be the assertion that N is divisible by some prime different from any of the primes $p_1, \ldots, p_n$. Now N is divisible by some prime p (if N is itself prime, then $p = N$), and N is not divisible by any of the primes $p_1, \ldots, p_n$, since each of these leaves a remainder of 1 when divided into N. Hence Q is true. Since Q is not compatible with the falsity of the theorem, the theorem is true. (Note that the assertion that N is divisible by some prime requires proof. This is easily given by induction; see Problem 3 of Section 1-3.)

Since we are momentarily concerned with logic, it might be helpful to say a word about implications. The assertion "P implies Q," where P and Q are sentences, means that Q can be derived from P by logically correct steps (more precisely: from P and the axioms of the system with which one is concerned). It says nothing about P and Q individually, but it makes a statement about a relationship between them. It can also be interpreted as meaning "whenever (or if) P is true, so is Q." It can be proved either by starting from P and deducing Q, or by starting from "Q is false" and deducing "P is false," the latter being an indirect proof. (See, for example, the inductive step in the proof that u_n and u_{n+1} have no common factor.) If P implies Q, then Q is said to be a *necessary* condition for P, since Q necessarily happens whenever P does, and P is said to be a *sufficient* condition for Q, since the truth of P guarantees (is sufficient for) that of Q. If P implies Q and Q implies P, then P and Q are said to be *equivalent* statements, each is said to be a *necessary and sufficient* condition for the other, and we say that one is valid *if and only if* the other is. For example, for a number larger than 2 to be prime, it is necessary, but not sufficient, that it be odd. In order that a polynomial assume both positive and negative values for appropriate values of x, it is sufficient, but not necessary, that it be of odd degree. It is a famous theorem of P. Dirichlet that there are infinitely many primes among the numbers m, $m + k$, $m + 2k$, $m + 3k, \ldots$, if and only if m and k have no common factor larger than 1. A necessary and sufficient condition for an integer to be divisible by 9 is that the sum of its digits be divisible by 9. A number is a square only if its final digit is one of 0, 1, 4, 6, or 9.

The above is a very brief introduction to the logic of mathematics, but it will suffice for this book. However, one more point should be made about the proofs encountered in elementary number theory, which verges more toward the psychological. It is a well-known phenomenon in mathematics that an excessively simple theorem frequently is difficult to prove (although the proof, in retrospect, may be short and elegant) just because of its simplicity. This is probably due in part to the lack of any hint in the statement of the theorem concerning the machinery to be used in proving it, and in part to the lack of available machinery. Many theorems of elementary number theory are of this kind, and there is considerable diversity in the types of arguments used in their proofs. When we are presented with a large number of theorems bearing on the same subject but proved by quite diverse means, the natural tendency is to regard the techniques used in the various proofs as special tricks, each applicable only to the theorem with which it is associated. *A technique ceases to be a trick and becomes a method only when it has been encountered enough times to seem natural;* correspondingly, a subject may be regarded as a "bag of tricks" if the ratio of techniques to results is too high. Unfortunately, elementary number theory has sometimes been regarded as such a subject. On working longer in the field, however, we find that many of the tricks become methods, and that there is more uniformity than is at first apparent. By making a conscious effort to abstract and retain the core of the proofs that follow, the reader will begin to see patterns emerging sooner than he otherwise might.

Consider, for example, the assertion that $\tau(n)$ is even unless n is a square, i.e., the square of another integer. This can be proved as follows: If d is a divisor of n, then so is the integer n/d. If n is not a square, then $d \neq n/d$, since otherwise $n = d^2$. Hence, if n is not a square, its divisors can be paired off into couples d, n/d, so that each divisor of n occurs just once as an element of some one of these couples. The number of divisors is therefore twice the number of couples and, being twice an integer, is even.

We have here applied the principle that in counting integers having a certain property (here "counting" may be replaced by "adding"), we may find it helpful first to group them in judicious fashion. There are several problems in the present book whose solutions depend on this idea.

Problems

1. Show that $\tau(n)$ is odd if n is a square.

2. Anticipating Theorem 1–1, suppose that every integer can be written in the form $6k + r$, where k is an integer and r is one of the numbers 0, 1, 2, 3, 4, 5. (a) Show that if $p = 6k + r$ is a prime different from 2 and 3, then $r = 1$ or 5. (b) Show that the product of numbers of the form $6k + 1$ is of the same form. (c) Show that there exists a prime of the form $6k - 1 = 6(k - 1) + 5$. (d) Show that there are infinitely many primes of the form $6k - 1$.

1–5 Radix representation. Although we have assumed a knowledge of the structure of the system of integers, we have said nothing about the method which will be used to assign names to the integers. There are, of course, various ways of doing this, of which the Roman and decimal systems are probably the best known. While the decimal system has obvious advantages over Roman numerals, and the advantage of familiarity over any other method, it is not always the best system for theoretical purposes. A rather more general scheme is sometimes convenient, and it is the object of the following two theorems to show that a representation of this kind is possible, i.e., that each integer is given a unique name. *Here, and throughout the remainder of the book, lower-case Latin letters will denote integers, except where otherwise specified.*

Theorem 1–1. If a is positive and b is arbitrary, there is exactly one pair of integers q, r such that the conditions

$$b = qa + r, \qquad 0 \le r < a, \tag{6}$$

hold.

Proof: First, we show that (6) has at least one solution.

Consider the set D of integers of the form $b - ua$, where u runs over all integers, positive and nonpositive. For the particular choice

$$u = \begin{cases} 0 & \text{if } b \ge 0, \\ b & \text{if } b < 0, \end{cases}$$

the number $b - ua$ is nonnegative, so that D contains nonnegative elements. The subset consisting of the nonnegative elements of D therefore has a smallest element. Take r to be this number, and q the value of u which corresponds to it; i.e., let q be the largest integer such that $b - qa \ge 0$. Then $r = b - qa \ge 0$, whereas

$$r - a = b - (q + 1)a < 0;$$

hence (6) is satisfied.

To show the uniqueness of q and r, assume that q' and r' also are integers such that

$$b = q'a + r', \qquad 0 \leq r' < a.$$

Then if $q' < q$, we have

$$b - q'a = r' \geq b - (q - 1)a = r + a \geq a,$$

and this contradicts the inequality $r' < a$. Hence $q' \geq q$. Similarly, we show that $q \geq q'$. Therefore $q = q'$, and consequently $r = r'$. ▲

THEOREM 1–2. Let g be greater than 1. Then each integer a greater than 0 can be represented uniquely in the form

$$a = c_0 + c_1 g + \cdots + c_n g^n,$$

where c_n is positive and $0 \leq c_m < g$ for $0 \leq m \leq n$.

Proof: We prove the representability by induction on a. For $a = 1$, we have $n = 0$ and $c_0 = 1$.

Take a greater than 1 and assume that the theorem is true for $1, 2, \ldots, a - 1$. Since g is larger than 1, the numbers $g^0, g^1, g^2, \ldots$ form an increasing sequence, and any positive integer lies between some pair of successive powers of g. More explicitly, there is a unique $n \geq 0$ such that $g^n \leq a < g^{n+1}$. By Theorem 1–1, there are unique integers c_n and r such that

$$a = c_n g^n + r, \qquad 0 \leq r < g^n.$$

Here $c_n > 0$, since $c_n g^n = a - r > g^n - g^n = 0$; moreover, $c_n < g$ because $c_n g^n \leq a < g^{n+1}$. If $r = 0$, then

$$a = 0 + 0 \cdot g + \cdots + 0 \cdot g^{n-1} + c_n g^n,$$

whereas if r is positive, the induction hypothesis shows that r has a representation of the form

$$r = b_0 + b_1 g + \cdots + b_t g^t,$$

where b_t is positive and $0 \leq b_m < g$ for $0 \leq m \leq t$. Moreover, $t < n$. Thus

$$a = b_0 + b_1 g + \cdots + b_t g^t + 0 \cdot g^{t+1} + \cdots + 0 \cdot g^{n-1} + c_n g^n,$$

where the terms with coefficient zero occur only if $t + 1 < n$. Now use the induction principle.

To prove uniqueness, assume that there are two distinct representations for a:

$$a = c_0 + c_1 g + \cdots + c_n g^n = d_0 + d_1 g + \cdots + d_r g^r,$$

with $n \geq 0$, $c_n > 0$, and $0 \leq c_m < g$ for $0 \leq m \leq n$, and also $r \geq 0$, $d_r > 0$, and $0 \leq d_m < g$ for $0 \leq m \leq r$. Then by subtracting one of these representations of a from the other, we obtain an equation of the form

$$0 = e_0 + e_1 g + \cdots + e_s g^s,$$

where s is the largest value of m for which $c_m \neq d_m$, so that $e_s \neq 0$. If $s = 0$, we have the contradiction $e_0 = e_s = 0$. If $s > 0$ we have

$$|e_m| = |c_m - d_m| \leq g - 1, \qquad 0 \leq m \leq s - 1,$$

and

$$e_s g^s = -(e_0 + \cdots + e_{s-1} g^{s-1}),$$

so that

$$\begin{aligned} g^s \leq |e_s g^s| = |e_0 + \cdots + e_{s-1} g^{s-1}| &\leq |e_0| + \cdots + |e_{s-1}| g^{s-1} \\ &\leq (g - 1)(1 + g + \cdots + g^{s-1}) = g^s - 1, \end{aligned}$$

which is also a contradiction. We conclude that $n = r$ and $c_m = d_m$ for $0 \leq m \leq n$, and the representation is unique. ▲

By means of Theorem 1–2 we can construct a system of names or symbols for the positive integers in the following way. We choose arbitrary symbols to stand for the digits (i.e., the nonnegative integers less than g) and replace the number

$$c_0 + c_1 g + \cdots + c_n g^n$$

by the simpler symbol $c_n c_{n-1} \ldots c_1 c_0$. For example, choosing g to be ten, and giving the smaller integers their customary symbols, we have the ordinary decimal system, in which, for example, 2743 is an abbreviation for the value of the polynomial $2x^3 + 7x^2 + 4x + 3$ when x is ten. But there is no reason why we must use ten as the base, or *radix*; if we used 7 instead, we would write the integer whose decimal representation is 2743 as 10666, since

$$2743 = 6 + 6 \cdot 7 + 6 \cdot 7^2 + 0 \cdot 7^3 + 1 \cdot 7^4.$$

To indicate the base being used, when it is different from ten, we might use a subscript (in the decimal system), so that for example,

$$2743 = (10666)_7.$$

Of course, if the radix is larger than 10, it will be necessary to invent symbols to replace $10, 11, \ldots, g - 1$. For example, taking $g = 12$ and setting $10 = \alpha$, $11 = \beta$, we have

$$(14)_{12} + (7)_{12} = (1\beta)_{12}$$

and

$$(31)_{12} \cdot (\alpha)_{12} = 37 \cdot 10 = 370 = (26\alpha)_{12}.$$

In addition to the usual base, 10, the numbers 2 and 12 have received serious attention as useful bases. The proponents of the base 12 (the *duodecimal* system, as it is called) argue that 12 is a better base than 10 because in the duodecimal system many more fractions have terminating decimal (or rather, duodecimal) expansions [e.g., $1/2 = (0.6)_{12}$, $1/3 = (0.4)_{12}$, $1/4 = (0.3)_{12}$, $1/6 = (0.2)_{12}$, $1/12 = (0.1)_{12}$], large numbers could be written in shorter form, and some systems of measurement (e.g., feet and inches) are already duodecimal. Be that as it may, and counter-arguments certainly exist, there does not seem to be the slightest chance of such a "reform" occurring, so the subject must remain in the realm of idle speculation.

The base 2 is another matter completely. The *binary* system, consisting of only two digits, 0 and 1, is in constant use today in the scientific world, specifically in modern high-speed computers; in these machines the two binary digits correspond to the physical alternative that something is or is not the case: current is or is not flowing, a spot on a magnetic tape is or is not magnetized, etc. If we liken digits to colors, we might say that in the binary system we can see only black and white, whereas in the decimal system the digits distinguish ten shades, from white through gray to black. In this sense a binary digit carries less information than a decimal digit, a fact reflected in the far greater number of binary digits required to express any number which is at all large; for example,

$$1024 = (10{,}000{,}000{,}000)_2.$$

The machine experts have neatly summarized this situation by abbreviating "binary digit" to "bit," indicating that one binary digit is one bit of information.

What may seem at first sight to be a disadvantage, namely that a large number of bits is required to represent a significant amount of information, is in fact only the other side of the coin of versatility; a bit is like a brick, in that it takes a lot of them to make anything interesting, but a very wide range of things can be made out of them, exactly because they have so little built-in structure.

PROBLEMS

1. (a) Show that using only the standard weights $1, 2, 2^2, \ldots, 2^n$, one can weigh any integral weight less than 2^{n+1} by putting the unknown weight on one pan of the balance and a suitable combination of standard weights on the other pan.

(b) Prove that no other set of $n + 1$ weights will do this. [*Hint:* Name the weights so that $w_0 \leq w_1 \leq \cdots \leq w_n$. Let k be the smallest index such that $w_k \neq 2^k$ and obtain a contradiction, using the fact that the number of nonempty subsets of a set of $n + 1$ elements is $2^{n+1} - 1$.]

2. Construct the addition and multiplication tables for the duodecimal digits, i.e., the digits in base 12. Using these tables, evaluate

$$(21\alpha9)_{12} \cdot (\beta370)_{12}.$$

3. To multiply two numbers, such as 37 and 22, set up a table according to the following pattern:

37	22
18	44
9	88
4	176
2	352
1	704

The first column is formed by successive halvings (fractional remainders are discarded whenever they occur) and the second by successive doublings. If the elements of the second column standing opposite odd numbers in the first are added together, the result is $22 + 88 + 704 = 814 = 22 \cdot 37$. Use the binary representation to show that this rule is general.

4. Let $u_1, u_2, \ldots$ be the Fibonacci sequence defined in Section 1–3.

(a) Prove by induction (or otherwise) that for $n > 0$,

$$u_{n-1} + u_{n-3} + u_{n-5} + \cdots < u_n,$$

the sum on the left continuing so long as the subscript remains larger than 1.

(b) Show that every positive integer can be represented in a unique way in the form $u_{n_1} + u_{n_2} + \cdots + u_{n_k}$, where $k \geq 1$, $n_{j-1} \geq n_j + 2$ for $j = 2, 3, \ldots, k$, and $n_k > 1$.

CHAPTER 2

THE EUCLIDEAN ALGORITHM AND ITS CONSEQUENCES

2–1 Divisibility. Let a be different from zero, and let b be arbitrary. Then, if there is a c such that $b = ac$, we say that a *divides* b, or that a is a *divisor* of b, and write $a|b$ (negation: $a \nmid b$). As usual, the letters involved represent integers.

The following statements are immediate consequences of this definition:

(1) For every $a \neq 0$, $a|0$ and $a|a$. For every b, $\pm 1|b$.
(2) If $a|b$ and $b|c$, then $a|c$.
(3) If $a|b$ and $a|c$, then $a|(bx + cy)$ for each x, y. (If $a|b$ and $a|c$, than a is said to be a *common divisor* of b and c.)
(4) If $a|b$ and $b \neq 0$, then $|a| \leq |b|$.

2–2 The Euclidean algorithm and greatest common divisor.

THEOREM 2–1. Given any two integers a and b not both zero, there is a unique integer d such that

(i) $d > 0$;
(ii) $d|a$ and $d|b$;
(iii) if d_1 is any integer such that $d_1|a$ and $d_1|b$, then $d_1|d$.

Property (iii) says that every common divisor of a and b divides d; from assertion (4) above, it follows that d is the numerically largest of the various divisors of a and b. Thus, among the common divisors of a and b, d is maximal in two different senses, and hence is called the *greatest common divisor* of a and b. We abbreviate this statement by saying that the GCD of a and b is d, and writing simply $(a, b) = d$. The nomenclature is somewhat misleading, because "greatest" seems to refer to size, whereas it is actually the maximality of d in the sense of (iii) which is important, and not its size.

Proof: First let a and b be positive, and suppose that a is the larger of the two numbers; otherwise we can simply interchange their names. By Theorem 1–1, there are unique integers q_1 and r_1 such that

$$a = bq_1 + r_1, \qquad 0 \leq r_1 < b.$$

If $r_1 = 0$, then b is a divisor of a and we can take $d = b$, insofar as conditions (i) through (iii) above are concerned: b is positive, it is a common divisor of a and b, and every common divisor of a and b is a divisor of b. We shall return to the question of uniqueness below.

If $r_1 \neq 0$, then repeated application of Theorem 1–1 shows the existence of unique pairs $q_2, r_2; q_3, r_3; \ldots$, such that

$$\begin{aligned}
a &= bq_1 + r_1, & 0 &< r_1 < b,\\
b &= r_1q_2 + r_2, & 0 &< r_2 < r_1,\\
r_1 &= r_2q_3 + r_3, & 0 &< r_3 < r_2,\\
&\vdots\\
r_{k-3} &= r_{k-2}q_{k-1} + r_{k-1}, & 0 &< r_{k-1} < r_{k-2},\\
r_{k-2} &= r_{k-1}q_k + r_k, & 0 &< r_k < r_{k-1},\\
r_{k-1} &= r_kq_{k+1}.
\end{aligned}$$

Here we are confronted at each stage with the possibility that the remainder is zero, but we have assumed that this does not happen until the kth stage, when we divide r_{k-1} by r_k; or, to put it the other way around, we define k as the number of the stage at which a zero remainder appears. The process must stop then, of course, since Theorem 1–1 does not provide for division by zero. On the other hand, a zero remainder must eventually occur, since each remainder is a nonnegative integer strictly smaller than the preceding one, and the existence of an infinite sequence of such numbers contradicts the well-ordering axiom. Thus if $b \nmid a$, there is always a finite system of equations of the kind above, and a last nonzero remainder r_k. We assert that *to satisfy conditions* (i) *through* (iii) *we can take* $d = r_k$. For from the last equation we see that $r_k | r_{k-1}$; from the preceding equation, using statement (2) of Section 2–1, we see that $r_k | r_{k-2}$, etc. Finally, from the second and first equations, respectively, it follows that $r_k | b$ and $r_k | a$. Thus r_k is a common divisor of a and b. Now let d_1 be any common divisor of a and b. From the first equation, $d_1 | r_1$; from the second, $d_1 | r_2$; etc.; from the next-to-last equation $d_1 | r_k$. Thus we can take the d of the theorem to be r_k.

If $a < b$, interchange the names of a and b. If either a or b is negative, find the d corresponding to $|a|$ and $|b|$. If a is zero, $(a, b) = |b|$.

If both d_1 and d_2 have the properties of the theorem, then d_1, being a common divisor of a and b, divides d_2. Similarly, $d_2 | d_1$. This clearly implies that $d_1 = d_2$, and the GCD is unique. ▲

The chain of operations indicated by the above equations is known as the *Euclidean algorithm;* as will be seen, it is a cornerstone of multiplicative number theory. (In general, an algorithm is a systematic procedure which is applied repeatedly, each step depending on the results of the earlier steps. Other examples are the long-division algorithm and the square-root algorithm.) The Euclidean algorithm is actually quite practicable in numerical cases; for example, if we wish to find the GCD of 4147 and 10672, we compute as follows:

$$\begin{aligned} 10672 &= 4147 \cdot 2 + 2378, \\ 4147 &= 2378 \cdot 1 + 1769, \\ 2378 &= 1769 \cdot 1 + 609, \\ 1769 &= 609 \cdot 2 + 551, \\ 609 &= 551 \cdot 1 + 58, \\ 551 &= 58 \cdot 9 + 29, \\ 58 &= 29 \cdot 2. \end{aligned}$$

Hence $(4147, 10672) = 29$.

It is frequently important to know whether two integers a and b have a common factor larger than 1. If they have not, so that $(a, b) = 1$, we say that they are *relatively prime*, or *prime to each other*.

The following properties of the GCD are easily derived either from the definition or from the Euclidean algorithm.

(a) The GCD of more than two numbers, defined as that positive common divisor which is divisible by every common divisor, exists and can be found in the following way. Let there be n numbers $a_1, a_2, \ldots, a_n$, and define

$$D_1 = (a_1, a_2), D_2 = (D_1, a_3), \ldots, D_{n-1} = (D_{n-2}, a_n).$$

Then $(a_1, a_2, \ldots, a_n) = D_{n-1}$.

(b) $(ma, mb) = m(a, b)$ if $m > 0$.

(c) If $m|a$ and $m|b$, then $(a/m, b/m) = (a, b)/m$, provided $m > 0$.

(d) If $(a, b) = d$, there exist integers x, y such that $ax + by = d$. [This last statement has an important consequence, namely, if a and b are relatively prime, there exist x, y such that $ax + by = 1$. Conversely, if there is such a representation of 1, then clearly $(a, b) = 1$.]

(e) If a given integer is relatively prime to each of several others, it is relatively prime to their product. For example, if $(a, b) = 1$ and $(a, c) = 1$, there are x, y, t, and u such that $ax + by = 1$ and $at + cu = 1$, whence

$$ax + by(at + cu) = a(x + byt) + bc(yu) = 1,$$

and therefore $(a, bc) = 1$.

The Euclidean algorithm can be used to find the x and y of property (d). Thus, using the numerical example above, we have

$$\begin{aligned}
29 &= 551 - 58 \cdot 9 && (58 = 609 - 551 \cdot 1)\\
&= 551 - 9(609 - 551 \cdot 1)\\
&= 10 \cdot 551 - 9 \cdot 609 && (551 = 1769 - 2 \cdot 609)\\
&= 10(1769 - 2 \cdot 609) - 9 \cdot 609\\
&= 10 \cdot 1769 - 29 \cdot 609 && (609 = 2378 - 1 \cdot 1769)\\
&= 10 \cdot 1769 - 29(2378 - 1 \cdot 1769)\\
&= 39 \cdot 1769 - 29 \cdot 2378 && (1769 = 4147 - 2378)\\
&= 39(4147 - 2378) - 29 \cdot 2378\\
&= 39 \cdot 4147 - 68 \cdot 2378 && (2378 = 10672 - 2 \cdot 4147)\\
&= 175 \cdot 4147 - 68 \cdot 10672,
\end{aligned}$$

so that $x = 175$, $y = -68$ is one pair of integers such that $4147x + 10672y = 29$. It is not the only such pair, as we shall see in Section 2–4.

Problems

1. Show that if $a|b$ and $b \neq 0$, then $|a| \leq |b|$.
2. Show that $(a, b) = (a, b + ka)$ for every k.
3. Show that if $(a, b) = 1$, then $(a - b, a + b) = 1$ or 2.
4. Show that if $ax + by = m$, then $(a, b)|m$.
5. Prove assertions (a) through (e) of the text.
6. (a) Evaluate (4655, 12075), and express the result as a linear combination of 4655 and 12075; that is, in the form $4655x + 12075y$. (b) Do the same for (1369, 2597). (c) Do the same for (2048, 1275).
7. Show that no cancellation is possible in the fraction

$$\frac{a_1 + a_2}{b_1 + b_2}$$

if $a_1b_2 - a_2b_1 = \pm 1$.

8. Evaluate the following:
(a) (493, 731, 1751); (b) (4410, 1404, 8712); (c) (703, 893, 1729, 33041).
9. Show that if $b|a$, $c|a$, and $(b, c) = 1$, then $bc|a$.
10. Show that if $(b, c) = 1$, then $(a, bc) = (a, b)(a, c)$. [*Hint:* Prove that each member of the alleged equation divides the other. Use property (d) in the text, and the preceding problem.]
11. In the notation introduced in the proof of Theorem 2–1, show that each nonzero remainder r_m, with $m \geq 2$, is less than $r_{m-2}/2$. [*Hint:* Consider separately the cases in which r_{m-1} is less than, equal to, or greater than $r_{m-2}/2$.] Deduce that the number of divisions in the Euclidean algorithm is at most $2n + 1$, where n is that integer such that $2^n \leq b < 2^{n+1}$, and where b is the smaller of the two numbers whose GCD is being found.

12. (a) Let D be the smallest positive number which can be represented in the form $ax + by$ with integers x and y. Show that if c is *any* integer representable in this form, then $D|c$. [*Hint:* Apply Theorem 1–1 and show that the remainder upon dividing c by D must be zero, because of the minimality of D.] (b) Show that $D|a$ and $D|b$. (c) Prove Theorem 2–1 without using the Euclidean algorithm.

13. Use the method of the preceding problem to prove the existence and uniqueness of an appropriately defined GCD of several integers $a_1, \ldots, a_n$, not all of which are zero.

14. Extend assertions (b) through (e) of the text to the case of several integers.

2–3 The unique factorization theorem.

THEOREM 2–2. Every integer $a > 1$ can be represented as a product of one or more primes. (It is customary to allow products to contain only one factor, and sums to contain only one term, since this simplifies the statements of theorems.)

Proof: The theorem is true for $a = 2$. Assume it to be true for 2, 3, 4, ..., $a - 1$. If a is prime, we are through. Otherwise a has a divisor different from 1 and a, and we have $a = bc$, with $1 < b < a$, $1 < c < a$. The induction hypothesis then implies that

$$b = \prod_{i=1}^{s} p'_i, \qquad c = \prod_{i=1}^{t} p''_i,$$

with p'_i, p''_i primes, and hence $a = p'_1 p'_2 \cdots p'_s p''_1 \cdots p''_t$. ▲

Any positive integer which is not prime and which is different from unity is said to be *composite.* Hereafter p will be used to denote a positive prime number, unless otherwise specified.

THEOREM 2–3. If $a|bc$ and $(a, b) = 1$, then $a|c$.

Proof: If $(a, b) = 1$, there are integers x and y such that $ax + by = 1$, or $acx + bcy = c$. But a divides both ac and bc, and hence the left side of this equation, and therefore a divides c. ▲

THEOREM 2–4. If

$$p \,\Big|\, \prod_{m=1}^{n} p_m,$$

then for at least one m, we have $p = p_m$.

Proof: Suppose that $p|p_1 p_2 \cdots p_n$ but that p is different from any of the primes $p_1, p_2, \ldots, p_{n-1}$. Then p is relatively prime to each of $p_1, \ldots, p_{n-1}$ and so, by property (e) of the preceding section, is relatively prime to their product. By Theorem 2–3, $p|p_n$, whence $p = p_n$. ▲

THEOREM 2–5. (*Unique Factorization Theorem*). The representation of $a > 1$ as a product of primes is unique up to the order of the factors.

Proof: We must show exactly the following. From

$$a = \prod_{m=1}^{n_1} p_m = \prod_{m=1}^{n_2} p'_m \quad (p_1 \leq p_2 \leq \cdots \leq p_{n_1}; p'_1 \leq p'_2 \leq \cdots \leq p'_{n_2}),$$

it follows that $n_1 = n_2$ and $p_m = p'_m$ for $1 \leq m \leq n_1$.

For $a = 2$ the assertion is true, since $n_1 = n_2 = 1$ and $p_1 = p'_1 = 2$. For $a > 2$, assuming the assertion to be correct for $2, 3, \ldots, a - 1$, we find:

(a) If a is prime, $n_1 = 1$, $p_1 = p'_1 = a$.
(b) Otherwise $n_1 > 1$, $n_2 > 1$. From

$$p'_1 \Big| \prod_{m=1}^{n_1} p_m, \qquad p_1 \Big| \prod_{m=1}^{n_2} p'_m$$

it follows by Theorem 2–4 that for at least one r and at least one s,

$$p'_1 = p_r, \qquad p_1 = p'_s.$$

Since

$$p_1 \leq p_r = p'_1 \leq p'_s = p_1,$$

we have $p_1 = p'_1$. Moreover, since $1 < p_1 < a$ and $p_1|a$, we have

$$1 < \frac{a}{p_1} = \prod_{m=2}^{n_1} p_m = \prod_{m=2}^{n_2} p'_m < a.$$

Thus the products $p_2p_3 \cdots p_{n_1}$ and $p'_2p'_3 \cdots p'_{n_2}$ are prime decompositions of the same number, and this number lies in the range in which, by the induction hypothesis, factorization is unique. Hence $n_1 = n_2$ and $p_1 = p'_1$, $p_2 = p'_2, \ldots$. Thus the two representations for a were identical. ▲

In view of its fundamental position in the theory of numbers, we give a second proof of the unique factorization theorem, this one being independent of the notion of GCD. As a preliminary step we note that if an integer n has the unique factorization property, and if a prime p divides n, then p actually occurs in the prime factorization of n; for otherwise we could write n/p as a product of primes, not necessarily unique, and multiplying through by p would yield a second representation for n as a product of primes.

Primes, by definition, have unique factorization; so let us consider an integer $n > 1$ which is not prime and let us suppose, as induction hypothesis, that all integers a with $1 < a < n$ have the unique factorization property. Suppose that n does not have it, and that we have the two representations

$$n = p_1 p_2 \cdots = p'_1 p'_2 \cdots,$$

where we again order the factors so that $p_1 \leq p_2 \leq \cdots$ and $p'_1 \leq p'_2 \leq \cdots$. We can suppose that no p_i is the same as any p'_i, since otherwise the common factor could be cancelled and the induction hypothesis applied. Because there are at least two factors in each representation, we have

$$n \geq p_1 p_2 \geq p_1^2 \quad \text{or} \quad p_1 \leq \sqrt{n},$$

and similarly $p'_1 \leq \sqrt{n}$. Thus the number $a = n - p_1 p'_1$ is nonnegative If a were 0, we would have

$$n = p_1 p'_1 = p_1 p_2 \cdots,$$
$$p'_1 = p_2 \cdots,$$

and hence $p'_1 = p_2$, contrary to our assumption. Therefore $a \geq 1$. But we also find that $a \neq 1$, since $a = 1$ would give $n = p_1 p'_1 + 1$, a number not divisible by p_1. Hence $a > 1$. By the induction hypothesis, a has unique factorization, and since both p_1 and p'_1 divide a, it follows from the preliminary remark that both of these primes must actually occur in the factorization of a. Furthermore, they are distinct, and consequently $a = p_1 p'_1 b$, where b is a positive integer. But then

$$n = a + p_1 p'_1 = p_1 p'_1 (b + 1) = p_1 p_2 \cdots,$$
$$p'_1 (b + 1) = p_2 \cdots,$$

and since $p_2 \cdots$ is a number with unique factorization and divisible by p'_1, it must be that p'_1 is one of the primes $p_2, \ldots$, contrary to our hypothesis. This contradiction shows that n has unique factorization, and it follows from the induction axiom that all integers larger than 1 have this property. ▲

At this point the question might well be raised, why all the fuss about a theorem whose truth seems perfectly obvious? The answer is, of course, that it seems obvious only because one is accustomed to it from experience with the small integers, and that one therefore *believes* that it is also true for larger integers. But believing and knowing are not the same thing.

It might be instructive to consider a situation rather similar to the one we have been concerned with, in which factorization is not unique. Instead of taking *all* the positive integers as our domain of discussion, suppose that we consider only those of the form $4k + 1$, namely $1, 5, 9, 13, \ldots$.

Call this set of integers D. The product of two elements of D is again in D, since

$$(4k + 1)(4m + 1) = 4(4km + k + m) + 1.$$

We could say that an element of D is *prime in D* if it is larger than 1 and has no factors in D except itself and 1; thus the first few numbers which are prime in D are 5, 9, 13, 17, 21, 29, ... It is now quite straightforward to show that every integer greater than 1 in D can be represented as a product of integers prime in D, but the unique factorization theorem does not hold, since, for example, 441 can be represented as products of numbers prime in D in two distinct ways: 21^2 and $9 \cdot 49$. The difficulty here is that D is not large enough, i.e., it does not contain the numbers 3 and 7, for example, which would be necessary to restore the unique factorization of 441. There is also no reason to suppose that the full set of integers is large enough, until it has been proved to be the case.

Problems

1. Show that if the reduced fraction a/b is a root of the equation

$$c_0x^n + c_1x^{n-1} + \cdots + c_n = 0,$$

where x is a real variable and $c_0, c_1, \ldots, c_n$ are integers with $c_0 \neq 0$, then $a|c_n$ and $b|c_0$. In particular, show that if k is an integer, then $\sqrt[n]{k}$ is rational if and only if it is an integer.

2. The unique factorization theorem shows that each integer $a > 1$ can be written uniquely as a product of powers of distinct primes. If the primes which do not divide a are included in this product with exponents 0, we can write

$$a = \prod_{i=1}^{\infty} p_i^{\alpha_i},$$

where p_i is the ith prime, $\alpha_i \geq 0$ for each i, $\alpha_i = 0$ for sufficiently large i, and the α_i's are uniquely determined by a. Show that if also

$$b = \prod_{i=1}^{\infty} p_i^{\beta_i},$$

then

$$(a, b) = \prod_{i=1}^{\infty} p_i^{\min(\alpha_i, \beta_i)},$$

where the symbol min (α, β) means the smaller of α and β. Use this fact to give different solutions to Problems 9 and 10, Section 2–2.

3. Show that the Diophantine equation

$$x^2 - y^2 = N$$

is solvable in nonnegative integers x and y if and only if N is odd or divisible by 4. Show further that the solution is unique if and only if $|N|$ or $|N|/4$, respectively, is unity or a prime. [*Hint:* Factor the left side.]

4. Show that every integer can be uniquely represented as the product of a square and a square-free number, the latter being an integer not divisible by the square of any prime.

5. Suppose that there are h primes not exceeding the positive integer x, so that $\pi(x) = h$. How many square-free numbers composed of one or more of these primes are there? How many squares not larger than x are there? Using the result of Problem 4, deduce that

$$\pi(x) \geq \frac{\log x}{2 \log 2}.$$

6. Show that the number $1 + 1/2 + 1/3 + \cdots + 1/n$ is not an integer for $n > 1$. [*Hint:* Consider the highest power of 2 occurring among $2, 3, \ldots, n$, and show that it occurs in only a single term.]

7. Suppose that $n = \prod_{i=1}^{r} p_i^{\alpha_i}$, where now the p_i are the primes actually dividing n, so that $\alpha_i > 0$ for $1 \leq i \leq r$. Show that every positive divisor of n is to be found exactly once among the terms resulting when the product

$$\prod_{i=1}^{r} (1 + p_i + \cdots + p_i^{\alpha_i})$$

is multiplied out. Deduce that the sum of the positive divisors of n is

$$\prod_{i=1}^{r} \frac{p_i^{\alpha_i+1} - 1}{p_i - 1},$$

and that the number of divisors of n is $\prod_{i=1}^{r} (\alpha_i + 1)$. Use the latter result to give a new proof that $\tau(n)$ is odd if and only if n is a square.

2–4 The linear Diophantine equation. For simplicity, we consider only the equation in two variables

$$ax + by = c. \tag{1}$$

It is easy to devise a scheme for finding an infinite number of solutions of this equation if any exist; the procedure can best be explained by means of a numerical example, say $5x + 22y = 18$. Since x is to be an integer, $\frac{1}{5}(18 - 22y)$ must also be integral. Writing

$$x = \frac{18 - 22y}{5} = 3 - 4y + \frac{3 - 2y}{5},$$

we see that $\frac{1}{5}(3 - 2y)$ must also be an integer, say z. This yields

$$z = \frac{3 - 2y}{5}, \quad \text{or} \quad 2y + 5z = 3.$$

We now repeat the argument, solving as before for the unknown which has the smaller coefficient:

$$y = \frac{3 - 5z}{2} = 1 - 2z + \frac{1 - z}{2},$$

$$\frac{1 - z}{2} = t, \qquad z = 1 - 2t.$$

Clearly, z will be an integer for any integer t, and we have

$$y = \frac{3 - 5(1 - 2t)}{2} = -1 + 5t,$$

$$x = \frac{18 - 22(-1 + 5t)}{5} = 8 - 22t.$$

What we have shown is that any solution x, y of the original equation must be of this form. By substitution, it is immediately seen that every pair of numbers of this form constitutes a solution, so that we have a general solution of the equation.

The same idea could be applied in the general case, but it is somewhat simpler to adopt a different approach. First of all, it should be noted that the left side of (1) is always divisible by (a, b), so that (1) has no solution unless $d|c$, where $d = (a, b)$. If this requirement is satisfied, we can divide through in (1) by d to obtain a new equation

$$a'x + b'y = c', \tag{2}$$

where $(a', b') = 1$. We now use property (d) of Section 2–2 to assert the existence of numbers x_0' and y_0' such that

$$a'x_0' + b'y_0' = 1,$$

whence $c'x_0'$, $c'y_0'$ is a solution of (2). We put $x_0 = c'x_0'$ and $y_0 = c'y_0'$.

Now suppose that x_1, y_1 is *any* solution of (2). We have

$$a'x_0 + b'y_0 = c',$$
$$a'x_1 + b'y_1 = c',$$

and, by subtraction,

$$a'(x_0 - x_1) = b'(y_1 - y_0).$$

Thus $a'|b'(y_1 - y_0)$, and since $(a', b') = 1$, it must be that $a'|(y_1 - y_0)$. Similarly, $b'|(x_0 - x_1)$, and since

$$\frac{a'}{b'} = \frac{y_1 - y_0}{x_0 - x_1},$$

it follows that there is an integer t such that

$$x_0 - x_1 = -b't,$$
$$y_0 - y_1 = a't,$$

or

$$x_1 = x_0 + b't,$$
$$y_1 = y_0 - a't.$$

Conversely, if x_1 and y_1 are related to a solution x_0, y_0 of (2) as in the equations just written, then

$$a'x_1 + b'y_1 = (a'x_0 + a'b't) + (b'y_0 - a'b't) = a'x_0 + b'y_0 = c',$$

and so x_1, y_1 is also a solution of (2). Since every solution of (1) is a solution of (2) and conversely, we have the following theorem.

THEOREM 2–6. A necessary and sufficient condition for the equation

$$ax + by = c$$

to have a solution x, y in integers is that $d|c$, where $d = (a, b)$. If there is one solution, there are infinitely many; they are exactly the numbers of the form

$$x = x_0 + \frac{b}{d}\,t, \qquad y = y_0 - \frac{a}{d}\,t,$$

where t is an arbitrary integer and x_0, y_0 is any one solution.

There are various ways of obtaining a particular solution. Sometimes one can be found by inspection; if not, the method explained at the beginning of the section may be used or, what is almost the same thing, the Euclidean algorithm may be applied to find a solution of the equation which results when the original equation is divided by (a, b). The latter process of successively eliminating the remainders in the Euclidean algorithm can be systematized, but we shall not do this here.

There are many "word problems" which lead to linear Diophantine equations that must be solved in *positive* integers, since only such solutions have meaning for the original problem. Suppose that the equation $ax + by = c$ is solvable in integers. Then we see from Theorem 2–6 that a positive solution will exist if and only if there is an integer t such that both

$$x_0 + \frac{b}{d}\,t > 0 \qquad \text{and} \qquad y_0 - \frac{a}{d}\,t > 0.$$

Let us first assume that a and b have opposite signs. Then the coefficients of t in the above inequalities have the same sign, and so either both require t to be not too small or both require t to be not too large; namely, if $b > 0$ and $a < 0$, we must have

$$t > -\frac{x_0 d}{b} \qquad \text{and} \qquad t > \frac{y_0 d}{a},$$

whereas if $b < 0$ and $a > 0$, we must have

$$t < -\frac{x_0 d}{b} \qquad \text{and} \qquad t < \frac{y_0 d}{a}.$$

In either case there is clearly an integer fulfilling the requirements, and in fact either all integers smaller than a certain one, or all integers larger than a certain one, will do. Hence, in this case, there are always infinitely many positive solutions of the equation.

The situation is quite different when a and b have the same sign. We can suppose that a and b are both positive, since otherwise we can multiply through in the original equation by -1. Then there is a positive solution if and only if there is an integer t such that

$$-\frac{x_0 d}{b} < t < \frac{y_0 d}{a},$$

and the number of positive solutions is the number of integers in this interval.

EXAMPLE. A sporting-goods store placed a total order of \$2490.00 for a number of bicycles at \$29 each and a number at \$33 each. How many bicycles of each kind were ordered?

We obtain the equation $29x + 33y = 2490$. From the Euclidean algorithm,

$$33 = 29 \cdot 1 + 4,$$
$$29 = 7 \cdot 4 + 1,$$

we have

$$1 = 29 - 7 \cdot 4 = 29 - 7(33 - 29) = 29 \cdot 8 - 33 \cdot 7,$$

and therefore a general solution of the equation is

$$x = 8 \cdot 2490 + 33t,$$
$$y = -7 \cdot 2490 - 29t.$$

The positive solutions correspond to integers t such that

$$-\frac{8 \cdot 2490}{33} < t < -\frac{7 \cdot 2490}{29},$$

or $-603.6\ldots < t < -601.03\ldots$. Thus there are two solutions, corresponding to $t = -602$ and $t = -603$:

$$x = 54, \qquad y = 28,$$

or

$$x = 21, \qquad y = 57.$$

Problems

1. Find a general solution of the linear Diophantine equation

$$2072x + 1813y = 2849.$$

2. Find all solutions of $19x + 20y = 1909$ with $x > 0$, $y > 0$.

*3. Let m and n be positive integers, with $m \leq n$, and let $x_0, x_1, \ldots, x_k$ be all the distinct numbers among the two sequences

$$\frac{0}{m}, \frac{1}{m}, \ldots, \frac{m}{m} \quad \text{and} \quad \frac{0}{n}, \frac{1}{n}, \ldots, \frac{n}{n},$$

arranged so that $x_0 < x_1 < \cdots < x_k$. Describe k as a function of m and n. What is the shortest distance between successive x's?

4. Let a and b be positive relatively prime integers. Then for certain nonnegative integers n (which we shall briefly refer to as the representable integers), the equation $ax + by = n$ has a solution with $x \geq 0$, $y \geq 0$, whereas for other n it does not have such a solution. For example, if $n = 0, 3, 5$, or 6, or if $n \geq 8$, then $3x + 5y = n$ has such a solution. Show that this example is typical, in the following sense: (a) There is always a number $N(a, b)$ such that for every $n \geq N(a, b)$, n is representable. (It may be helpful to combine the theory of the present section with the elementary analytic geometry of the line $ax + by = c$, interpreting x and y in the latter case as real variables. Note that so far it is only the existence of $N(a, b)$ which is in question, and not its size.) *(b) The minimal value of $N(a, b)$ is always $(a - 1)(b - 1)$. *(c) Exactly half the integers up to $(a - 1)(b - 1)$ are representable.

5. Apply the method discussed in the text, of repeatedly solving for the unknown with smallest coefficient, to solve the equation $1321x + 5837y + 1926z = 2983$.

6. Find necessary and sufficient conditions that the Diophantine equation $a_1x_1 + \cdots + a_nx_n = b$ should have an integral solution.

7. When Mr. Smith cashed a check for x dollars and y cents, he received instead y dollars and x cents, and found that he had two cents more than twice the proper amount. For how much was the check written?

2–5 The least common multiple.

THEOREM 2–7. The number

$$\langle a, b\rangle = \frac{|ab|}{(a, b)}$$

has the following properties: (1) $\langle a, b\rangle \geq 0$; (2) $a|\langle a, b\rangle$ and $b|\langle a, b\rangle$; (3) If $a|m$ and $b|m$, then $\langle a, b\rangle|m$.

Proof: (1) Obvious.

(2) Since $(a, b)|b$, we can write

$$\langle a, b\rangle = |a| \cdot \frac{|b|}{(a, b)},$$

and hence $a|\langle a, b\rangle$. Similarly,

$$\langle a, b\rangle = |b| \cdot \frac{|a|}{(a, b)},$$

and so $b|\langle a, b\rangle$.

(3) Let $m = ra = sb$, and set

$$d = (a, b), \qquad a = a_1 d, \qquad b = b_1 d.$$

Then

$$m = ra_1 d = sb_1 d;$$

thus $a_1|sb_1$, and since $(a_1, b_1) = 1$, we must have $a_1|s$. Thus $s = a_1 t$, and

$$m = ta_1 b_1 d = t\,\frac{ab}{d}. \ \blacktriangle$$

Because of the properties listed in Theorem 2–7, the number $\langle a, b\rangle$ is called the *least common multiple* (LCM) of a and b. The definition is easily extended to the case of more than two numbers, just as for the GCD. It is useful to remember that

$$ab = \pm(a, b)\langle a, b\rangle.$$

PROBLEMS

1. In the notation of Problem 2, Section 2–3, show that

$$\langle a, b\rangle = \prod_{i=1}^{\infty} p_i^{\max(\alpha_i, \beta_i)},$$

where max (α, β) is the larger of α and β. Use this to give a second proof of part (3) of Theorem 2–7.

2. Show that

$$\min\big(\alpha, \max(\beta, \gamma)\big) = \max\big(\min(\alpha, \beta), \min(\alpha, \gamma)\big).$$

(By symmetry, one may suppose that $\beta \geq \gamma$.) Deduce that

$$(a, \langle b, c\rangle) = \langle (a, b), (a, c)\rangle.$$

CHAPTER 3

CONGRUENCES

3-1 Introduction. The problem of solving the Diophantine equation $ax + by = c$ is that of finding an x such that ax and c leave the same remainder when divided by b, since then $b|(c - ax)$, and we can take $y = (c - ax)/b$. As we shall see, there are also many other instances in which a comparison must be made of the remainders after dividing each of two numbers a and b by a third, say m. Of course, if the remainders are the same, then $m|(a - b)$, and conversely, and this might seem to be an adequate notation. But as Gauss noticed, for most purposes the following notation is more suggestive: if $m|(a - b)$, then we write $a \equiv b \pmod{m}$, and say that *a is congruent to b modulo m.* (This has nothing to do with geometric congruence, of course.)

The use of the symbol "$\equiv$" is suggested by the similarity of the relation we are discussing to ordinary equality. Each of these two relations is an example of an *equivalence relation*, i.e., of a relation $\mathcal{R}$ between elements of a set, such that if a and b are arbitrary elements, either a stands in the relation $\mathcal{R}$ to b (more briefly, $a\mathcal{R}b$) or it does not, and which furthermore has the following properties:

(a) $a\mathcal{R}a$.
(b) If $a\mathcal{R}b$, then $b\mathcal{R}a$.
(c) If $a\mathcal{R}b$ and $b\mathcal{R}c$, then $a\mathcal{R}c$.

These are called the reflexive, symmetric, and transitive properties, respectively. That equality between numbers is an equivalence relation is obvious (or it may be taken as an axiom): either $a = b$ or $a \neq b$; $a = a$; if $a = b$, then $b = a$; if $a = b$ and $b = c$, then $a = c$.

THEOREM 3-1. Congruence modulo a fixed number m is an equivalence relation.

Proof:

(a) $m|(a - a)$, so that $a \equiv a \pmod{m}$.

(b) If $m|(a - b)$, then $m|(b - a)$; thus if $a \equiv b \pmod{m}$, then $b \equiv a \pmod{m}$.

(c) If $m|(a - b)$ and $m|(b - c)$, then $a - b \equiv km$, $b - c = lm$, say, so that $a - c = (k + l)m$; thus if $a \equiv b \pmod{m}$ and $b \equiv c \pmod{m}$, then $a \equiv c \pmod{m}$. ▲

Since the student will have occasion later to use other equivalence relations, we pause to show a simple but important property common to all such relations. If $\mathcal{R}$ is an equivalence relation with respect to a set S, then corresponding to each element a of S there is a subset S_a of S which consists of exactly those elements of S which are equivalent to a, so that b is in S_a if and only if $a\mathcal{R}b$. Now if $a\mathcal{R}b$, then the sets S_a and S_b are identical: if c is in S_b, then $c\mathcal{R}b$, and since $a\mathcal{R}b$, also $c\mathcal{R}a$, so that c is in S_a. If, on the other hand, a is not equivalent to b, then S_a and S_b are disjoint; that is, they have no element in common. For if c is in S_a and in S_b, then $c\mathcal{R}a$ and $c\mathcal{R}b$, which entails $a\mathcal{R}b$. These disjoint sets S_a, which together make up S, are called *equivalence classes;* an element of an equivalence class is sometimes called a *representative* of the class, and a *complete system of representatives* is any subset of S which contains exactly one element from each equivalence class.

Section 3-3 provides examples of all these notions, with somewhat different terminology.

PROBLEMS

1. Decide whether each of the following is an equivalence relation. If it is, describe the equivalence classes.
   (a) Congruence of triangles.
   (b) Similarity of triangles.
   (c) The relations "$\neq$", "$>$", and "$\geq$", relating real numbers.
   (d) Parallelism of lines.
   (e) Having the same mother.
   (f) Having a parent in common.
2. Define the relation $\mathcal{R}$ by $a\mathcal{R}b$ if and only if $a|b$. Show that $\mathcal{R}$ is reflexive and transitive, but not symmetric. Find other mathematically defined relations to show that any one or two of the properties of reflexivity, symmetry, and transitivity may hold without the others.
3. Show that if $a \equiv b \pmod{m}$ and $d|m$, then $a \equiv b \pmod{d}$.

3-2 Elementary properties of congruences. One reason for the superiority of the congruence notation is that congruences can be combined in much the same way as can equations.

THEOREM 3-2. *If $a \equiv b \pmod{m}$ and $c \equiv d \pmod{m}$, then $a + c \equiv b + d \pmod{m}$, $ac \equiv bd \pmod{m}$, and $ka \equiv kb \pmod{m}$ for every integer k.*

Proof: These statements follow immediately from the definition. For if $a \equiv b \pmod{m}$, then $m|(a - b)$ and, similarly, $m|(c - d)$, and therefore $m|(a - b + c - d)$, or $m|((a + c) - (b + d))$. But this means that

$a + c \equiv b + d \pmod{m}$. Secondly, if $m|(a - b)$ and $m|(c - d)$, then $m|(a - b)(c - d)$. But

$$(a - b)(c - d) = ac - bd + b(d - c) + d(b - a),$$

and since m divides the second and third terms on the right-hand side, also $m|(ac - bd)$. Finally, if $m|(a - b)$, then also $m|k(a - b)$ for every k. ▲

The situation is a little more complicated when we consider dividing both sides of a congruence by an integer. We cannot deduce from $ka \equiv kb \pmod{m}$ that $a \equiv b \pmod{m}$, for it may be that part of the divisibility of $ka - kb = k(a - b)$ by m is accounted for by the presence of the factor k. What is clearly necessary is that the part of m which does not divide k should divide $a - b$.

THEOREM 3–3. If $ka \equiv kb \pmod{m}$ and $(k, m) = d$, then

$$a \equiv b \bmod \left(\frac{m}{d}\right).$$

Proof: Theorem 2–3. ▲

THEOREM 3–4. If $f(x)$ is a polynomial with integral coefficients, and $a \equiv b \pmod{m}$, then $f(a) \equiv f(b) \pmod{m}$.

Proof: Let $f(x) = c_0 + c_1x + \cdots + c_nx^n$. If $a \equiv b \pmod{m}$, then for every nonnegative integer j,

$$a^j \equiv b^j \pmod{m},$$

and

$$c_ja^j \equiv c_jb^j \pmod{m},$$

by Theorem 3–2. Adding these last congruences for $j = 0, 1, \ldots, n$, we have the theorem. ▲

Theorem 3–4 is basic to much of what follows in this chapter. As a first very simple application of it, let us consider the well-known rule that a number is divisible by 9 if and only if the sum of the digits in its decimal expansion is divisible by 9. If for example $n = 3{,}574{,}856$, then $3 + 5 + 7 + 4 + 8 + 5 + 6 = 38$, and since 38 is not divisible by 9, neither is n. Here

$$n = 3 \cdot 10^6 + 5 \cdot 10^5 + 7 \cdot 10^4 + 4 \cdot 10^3 + 8 \cdot 10^2 + 5 \cdot 10 + 6,$$

so that $n = f(10)$, where

$$f(x) = 3x^6 + 5x^5 + 7x^4 + 4x^3 + 8x^2 + 5x + 6.$$

On the other hand, $f(1)$ is exactly the sum of the digits:

$$f(1) = 3 + 5 + 7 + 4 + 8 + 5 + 6.$$

Since $10 \equiv 1 \pmod 9$, it follows from Theorem 3–3 that also $f(10) \equiv f(1) \pmod 9$, and this implies in particular that either $f(10)$ and $f(1)$ both are divisible by 9 or neither is.

The same argument applies in general. The decimal representation of n is always the expression of n as the value of a certain polynomial $f(x)$ for $x = 10$, and invariably $f(10) \equiv f(1) \pmod 9$. We see in fact that the rule can be strengthened in the following way: if $n = f(10)$ and $m = g(10)$, then

$$n + m = f(10) + g(10) \equiv f(1) + g(1) \pmod 9,$$
$$n \cdot m = f(10) \cdot g(10) \equiv f(1) \cdot g(1) \pmod 9;$$

hence, if $n + m = F(10)$ and $n \cdot m = G(10)$, then

$$F(10) \equiv F(1) \equiv f(1) + g(1) \pmod 9,$$
$$G(10) \equiv G(1) \equiv f(1) \cdot g(1) \pmod 9.$$

In words, these last two congruences say the following: *The sum of the digits in $n + m$ is congruent* (mod 9) *to the sum of all the digits in n and m, and the sum of the digits in $n \cdot m$ is congruent* (mod 9) *to the product of the sum of the digits in n and the sum of the digits in m.* This statement provides a partial check on the correctness of arithmetical operations, called "casting out nines," which amounts simply to verifying that the italicized assertion holds in particular cases. If, for example, we computed $47 + 94$ as 131, we could recognize the existence of an error by noting that $(4 + 7) + (9 + 4) = 24 \equiv 6 \pmod 9$, whereas $1 + 3 + 1 \equiv 5 \pmod 9$. Similarly, it cannot be that $47 \cdot 19 = 793$, since $(4 + 7)(1 + 9) = 110 \equiv 1 + 1 + 0 \equiv 2 \pmod 9$, while $7 + 9 + 3 = 19 \equiv 1 \pmod 9$. On the other hand, it is also true that $47 \cdot 19 \neq 884$, even though $8 + 8 + 4 \equiv 2 \pmod 9$; hence this method does not afford an absolute check on accuracy.

Problems

1. Let
$$f(x) = a_0x^n + a_1x^{n-1} + \cdots + a_n,$$
where $a_0, \ldots, a_n$ are integers. Show that if d consecutive values of f (i.e., values for consecutive integers) are all divisible by the integer d, then $d \mid f(x)$ for all integral x. Show by an example that this sometimes happens with $d > 1$ even when $(a_0, \ldots, a_n) = 1$.

2. Using the fact that $10 \equiv -1 \pmod{11}$, devise a test for divisibility of an integer by 11, in terms of properties of its digits.

3. Use the fact that $7 \cdot 11 \cdot 13 = 1001$ to obtain a test for divisibility by any of the integers 7, 11, or 13.

4. Without carrying out the computations, test the accuracy of the following equations:

$$\text{(a) } 1097 \times 8156 = 8947132, \qquad \text{(b) } 28^3 + 37^3 = 73605.$$

5. Show that no square has a decimal expansion ending in 79. More generally, find all possible two-digit endings for squares.

6. Show that every square is congruent to 0 or 1 (mod 8). Deduce that no integer of the form $8k + 7$ is the sum of the squares of three integers.

7. Show that for every x, $x^3 \equiv x \pmod 3$, and that $x^5 \equiv x \pmod 5$. Formulate a general conjecture, and test it in some other cases.

8. Show that every quadratic discriminant $b^2 - 4ac$ is congruent to 0 or 1 (mod 4).

9. Show that if $(x, 6) = 1$, then $x^2 \equiv 1 \pmod{24}$.

10. Show that if $a \equiv b \pmod m$, then $(a, m) = (b, m)$.

3–3 Residue classes and arithmetic (mod m). When dealing with congruences modulo a fixed integer m, the set of all integers breaks down into m classes, called the *residue classes* (mod m), such that any two elements of the same class are congruent and two elements from two different classes are incongruent. The residue classes are also called arithmetic progressions with difference m. For many purposes it is completely immaterial which element of one of these residue classes is used; for example, Theorem 3–4 shows this to be the case when one considers the values modulo m of a polynomial with integral coefficients. In these instances it suffices to consider an arbitrary set of representatives of the various residue classes; that is, a set consisting of one element of each residue class. Such a set $a_1, a_2, \ldots, a_m$, called a *complete residue system modulo m*, is characterized by the following properties.

(a) If $i \neq j$, then $a_i \not\equiv a_j \pmod m$.
(b) If a is any integer, there is an index i with $1 \leq i \leq m$ for which $a \equiv a_i \pmod m$.

Examples of complete residue systems (mod m) are the set of integers $0, 1, 2, \ldots, m - 1$ and the set $1, 2, \ldots, m$. The elements of a complete residue system need not be consecutive integers, however; for $m = 5$ we could take $1, 22, 13, -6, 2500$, for example. More generally, if we write out the five arithmetic progressions with difference 5:

$$\begin{array}{l} \ldots, -10, -5, 0, 5, 10, 15, \ldots, \\ \ldots, -9, -4, 1, 6, 11, 16, \ldots, \\ \ldots, -8, -3, 2, 7, 12, 17, \ldots, \\ \ldots, -7, -2, 3, 8, 13, 18, \ldots, \\ \ldots, -6, -1, 4, 9, 14, 19, \ldots, \end{array}$$

we could choose any one element from each row, that from the first row being representative of all the integers divisible by 5, that from the second row being representative of all the integers of the form $5n + 1$, that from the third row being representative of all the integers of the form $5n + 2$, etc.

THEOREM 3–5. If $a_1, a_2, \ldots, a_m$ is a complete residue system (mod m) and $(k, m) = 1$, then $ka_1, ka_2, \ldots, ka_m$ also is a complete residue system (mod m).

Proof: We show directly that properties (a) and (b) above hold for this new set.

(a) If $ka_i \equiv ka_j \pmod{m}$, then by Theorem 3–3, $a_i \equiv a_j \pmod{m}$, whence $i = j$.

(b) Theorem 2–6 shows that if $(k, m) = 1$, the congruence $kx \equiv a \pmod{m}$ has a solution for any fixed a. Let a solution be x_0. Since $a_1, \ldots, a_m$ is a complete residue system, there is an index i such that $x_0 \equiv a_i \pmod{m}$. Hence $kx_0 \equiv ka_i \equiv a \pmod{m}$. ▲

When we restrict ourselves to a particular residue system (mod m), say $0, 1, \ldots, m - 1$, we obtain the "arithmetic (mod m)" if we work out the addition and multiplication tables for these m numbers. If, for example, we take $m = 5$, we obtain the following tables:

TABLE 3–1

(a)

+	0	1	2	3	4
0	0	1	2	3	4
1	1	2	3	4	0
2	2	3	4	0	1
3	3	4	0	1	2
4	4	0	1	2	3

(b)

×	0	1	2	3	4
0	0	0	0	0	0
1	0	1	2	3	4
2	0	2	4	1	3
3	0	3	1	4	2
4	0	4	3	2	1

In the first table, the entry in the row beginning with r and the column beginning with s is the sum $r + s$, in the sense that it is the representative of the residue class (mod 5) containing that sum. Thus $3 + 3 = 6 \equiv 1 \pmod{5}$, and this is the 1 in the next-to-last row and column of the first table. On the other hand, $3 \cdot 3 = 9 \equiv 4 \pmod{5}$, and 4 is the entry in the next-to-last row and column of the second table. This "modular" multiplication is perhaps new to the student, but "modular" addition is familiar to everyone through our systems of keeping time. When we say, "It is now seven o'clock; in 8 hours it will be three o'clock," we are simply adding modulo 12: $7 + 8 = 15 \equiv 3 \pmod{12}$. Similarly, the statement,

"Five days from next Thursday will be a Tuesday," entails addition (mod 7).

In the special case $m = 5$ it is possible to perform not only addition and multiplication but also subtraction and division, except for division by zero. In general, to subtract a from b means "find x such that $a + x$ is b." In ordinary arithmetic the word "is" in the quoted sentence means "is equal to," whereas in arithmetic (mod m) it must be taken to mean "is congruent to, modulo m." With this meaning we can verify that subtraction (mod m) is always possible by noting that in the addition table (Table 3–1a), each row in the body of the table contains all of the numbers 0, 1, 2, 3, 4, and each just once. To subtract 3 from 2 or to find what must be added to 3 to yield 2, we look along the row headed 3 until we encounter the 2, and obtain the number at the head of the column containing it, namely 4, as the difference: $2 - 3 \equiv 4 \pmod 5$. Division is carried out in the same way in Table 3–1(b); being able to do so depends on the fact that, excluding the first row and column in the body of the table, each of the numbers 1, 2, 3, 4 occurs exactly once in each row. Here we have interpreted the division of b by a as the finding of an x such that $b \equiv ax \pmod 5$.

With respect to division, a composite modulus is somewhat less satisfactory, because the fundamental principle is no longer valid that a product is zero only if one of the factors is zero. For example, $2 \cdot 3 \equiv 0 \pmod 6$, even though neither 2 nor 3 is 0 (mod 6). This situation is reflected in the fact that division is not always possible, since, for example, there is no sense to be attached to the symbol $1/2 \pmod 6$ because there is no x for which $2x \equiv 1 \pmod 6$. We shall return to this question in Section 3–5.

Problems

1. Let $m > 1$ be fixed. Show that if the integers $a_1, a_2, \ldots, a_k$ have any two of the following three properties, they also have the third, and hence constitute a complete residue system (mod m):

(a) If $i \neq j$, then $a_i \not\equiv a_j \pmod m$;

(b) if a is any integer, there is an index i with $1 \leq i \leq k$ for which $a \equiv a_k \pmod m$;

(c) $k = m$.

Prove Theorem 3–5 by verifying (a) and (c), rather than (a) and (b) as is done in the text.

2. Prove a theorem similar to Theorem 3–5, concerning the numbers $ka_1 + l$, $ka_2 + l, \ldots, ka_m + l$, in which l is any fixed integer.

3–4 Reduced residue systems and Euler's φ-function. The reason that we use the adjective "complete" when speaking of a residue system is that there is another kind which is also frequently useful, called a *reduced residue system*. This is a set of integers $a_1, \ldots, a_h$, incongruent (mod m) and relatively prime to m, such that if a is any integer prime to m, there is an index i, $1 \leq i \leq h$, for which $a \equiv a_i \pmod{m}$. In other words, a reduced residue system is a set of representatives, one from each of the residue classes containing integers prime to m. [Clearly, $(a, m) = (b, m)$ if $a \equiv b \pmod{m}$. For if a and b are congruent (mod m), then $m|(a - b)$, and since $(a, m)|m$, we have $(a, m)|(a - b)$. It follows that $(a, m)|b$, and consequently that $(a, m)|(b, m)$. By similar reasoning, $(b, m)|(a, m)$, and therefore $(a, m) = (b, m)$.] For example, 1 and 5 constitute a reduced residue system (mod 6), and 1, 2, 3, 4, 5, 6 a reduced residue system (mod 7). In the case of prime modulus p, a reduced residue system results from a complete residue system by omission of the single number divisible by p.

The number h of elements in a reduced residue system (mod m) is the number of positive integers not exceeding m and prime to m. This quantity, which depends on m, is customarily designated by $\varphi(m)$, and is called Euler's φ-function, after the Swiss mathematician Leonard Euler. It might be mentioned that for $m > 1$, $\varphi(m)$ can also be defined as the number of positive integers *less than* m and prime to m, since for such m, $(m, m) > 1$. For $m = 1$, however, the two definitions give different values.

THEOREM 3–6. *If $a_1, \ldots, a_{\varphi(m)}$ is a reduced residue system (mod m) and $(k, m) = 1$, then also $ka_1, \ldots, ka_{\varphi(m)}$ is a reduced residue system (mod m).*

The proof is exactly parallel to that of Theorem 3–5.

Table 3–2 lists the first few values of $\varphi(m)$:

TABLE 3–2

m	$\varphi(m)$	m	$\varphi(m)$	m	$\varphi(m)$
1	1	11	10	21	12
2	1	12	4	22	10
3	2	13	12	23	22
4	2	14	6	24	8
5	4	15	8	25	20
6	2	16	8	26	12
7	6	17	16	27	18
8	4	18	6	28	12
9	6	19	18	29	28
10	4	20	8	30	8

One immediately notices that for $m > 2$, the values of $\varphi(m)$ are even. This is always the case, since if a is one of the integers counted in $\varphi(m)$, that is, one of the integers not larger than m and prime to m, then $m - a$ is another such integer [for clearly $(a, m) = (m - a, m)$]. The two integers a and $m - a$ are distinct, since $a = m - a$ gives $m = 2a$, which is inconsistent with the assumption that $(m, a) = 1$, unless $a = 1$, $m = 2$. Hence, for $m > 2$, the integers counted in $\varphi(m)$ can be paired off, and so the number of them must be even.

Aside from the evenness of $\varphi(m)$, and the fact that $\varphi(p) = p - 1$ if p is a prime, the values of the φ-function seem to be highly irregular. However, as we shall soon see, it is possible to compute the value of $\varphi(m)$ very quickly if the prime factorization of m is known. The φ-function has many interesting properties, and it occurs repeatedly in number-theoretic investigations.

One feature to note from the above table is that in certain cases at least, the values $\varphi(m)$ and $\varphi(n)$ can be multiplied together to give $\varphi(mn)$; for example, $\varphi(3)\varphi(7) = \varphi(21)$, and $\varphi(4)\varphi(5) = \varphi(20)$. On the other hand, $\varphi(4)\varphi(6) \neq \varphi(24)$. The correct rule is as follows:

THEOREM 3–7. If $(m, n) = 1$, then $\varphi(mn) = \varphi(m)\varphi(n)$.

(A function with this property is called a *multiplicative function*. For another example, see Problem 10, Section 2–2.)

Proof: Take integers m, n with $(m, n) = 1$, and consider the numbers of the form $mx + ny$. If we can so restrict the values which x and y assume that these numbers form a reduced residue system (mod mn), there must be $\varphi(mn)$ of them. But their number is also the product of the number of values which x assumes and the number of values which y assumes. Clearly, in order for $mx + ny$ to be prime to m, it is necessary that $(m, y) = 1$, and likewise we must have $(n, x) = 1$. Conversely, if these last two conditions are satisfied, then $(mx + ny, mn) = 1$, since in this case any prime divisor of m, or of n, divides exactly one of the two terms in $mx + ny$. Hence let x range over a reduced residue system (mod n), say $x_1, \ldots, x_{\varphi(n)}$, and let y run over a reduced residue system (mod m), say $y_1, \ldots, y_{\varphi(m)}$. If for some indices i, j, k, l we have

$$mx_i + ny_j \equiv mx_k + ny_l \pmod{mn},$$

then

$$m(x_i - x_k) + n(y_j - y_l) \equiv 0 \pmod{mn}.$$

Since divisibility by mn implies divisibility by m, we have

$$\begin{aligned} m(x_i - x_k) + n(y_j - y_l) &\equiv 0 \pmod{m}, \\ n(y_j - y_l) &\equiv 0 \pmod{m}, \\ y_j &\equiv y_l \pmod{m}, \end{aligned}$$

whence $j = l$. Similarly, $i = k$. Thus the numbers $mx + ny$ so formed are incongruent (mod mn). Now let a be any integer prime to mn; in particular, $(a, m) = 1$ and $(a, n) = 1$. Then Theorem 2–6 shows that there are integers X, Y (not necessarily in the chosen reduced residue systems) such that $mX + nY = a$, whence also $mX + nY \equiv a \pmod{mn}$. Since $(m, Y) = (n, X) = 1$, there is an x_i such that $X \equiv x_i \pmod{n}$, and there is a y_j such that $Y \equiv y_j \pmod{m}$. This means that there are integers k, l such that $X = x_i + kn$, $Y = y_j + lm$. Therefore

$$mX + nY = m(x_i + kn) + n(y_j + lm) \equiv mx_i + ny_j \equiv a \pmod{mn}.$$

Hence, as x and y run over fixed reduced residue systems (mod n) and (mod m), respectively, $mx + ny$ runs over a reduced residue system (mod mn), and the proof is complete. ▲

THEOREM 3–8.

$$\varphi(m) = m \prod_{p|m} \left(1 - \frac{1}{p}\right),$$

where the notation indicates a product over all the distinct primes which divide m.

Proof: By Theorem 3–7, if

$$m = \prod_{i=1}^{r} p_i^{\alpha_i},$$

then

$$\varphi(m) = \prod_{i=1}^{r} \varphi(p_i^{\alpha_i}).$$

But we can easily evaluate $\varphi(p^\alpha)$ directly: all the positive integers not exceeding p^α are prime to p^α except the multiples of p, and there are just $p^{\alpha-1}$ of these, so that

$$\varphi(p_i^{\alpha_i}) = p_i^{\alpha_i} - p_i^{\alpha_i - 1} = p_i^{\alpha_i}\left(1 - \frac{1}{p_i}\right).$$

Thus

$$\varphi(m) = \prod_{i=1}^{r} p_i^{\alpha_i}\left(1 - \frac{1}{p_i}\right) = \prod_{i=1}^{r} p_i^{\alpha_i} \cdot \prod_{i=1}^{r}\left(1 - \frac{1}{p_i}\right)$$

$$= m \prod_{p|m}\left(1 - \frac{1}{p}\right). \quad ▲$$

For example, the four integers 1, 5, 7, 11 are all those which do not exceed 12 and are prime to 12, and

$$\varphi(12) = 12(1 - \tfrac{1}{2})(1 - \tfrac{1}{3}) = 4.$$

In Theorem 3–8 we have an example of the Π-symbol used for a product in which the variable index does not run over all the integers up to a certain one, but over the integers satisfying certain conditions. Whenever the range of summation or multiplication consists of anything more complicated than all the integers of a certain interval, the description of the range is written entirely below the $\sum$ or Π. Further examples occur in the proof of the next theorem. Perhaps it should also be mentioned that the symbol $\sum_{\cdots} 1$ means to add as many 1's as there are integers satisfying the conditions occurring below the $\sum$; in other words, it is the number of integers satisfying these conditions.

THEOREM 3–9.

$$\sum_{d|n} \varphi(d) = n.$$

Proof: Let $d_1, \ldots, d_k$ be the positive divisors of n. We separate the integers between 1 and n inclusive into classes $C(d_1), \ldots, C(d_k)$, putting an integer into the class $C(d_i)$ if its GCD with n is d_i. The number of elements in $C(d_i)$ is then

$$\sum_{\substack{a \leq n \\ (a,n)=d_i}} 1,$$

and since every integer up to n is in exactly one of the classes, we have

$$\sum_{d_i|n} \sum_{\substack{a \leq n \\ (a,n)=d_i}} 1 = n.$$

The number of integers a such that $1 \leq a \leq n$ and $(a, n) = d_i$ is exactly equal to the number of integers b such that $1 \leq b \leq n/d_i$ and $(b, n/d_i) = 1$; in fact, multiplying the b's by d_i, we obtain the a's. But from the definition of the Euler function, the number of b's is clearly $\varphi(n/d_i)$. Thus

$$\sum_{d_i|n} \varphi\left(\frac{n}{d_i}\right) = n,$$

which is equivalent to the theorem, since, as d_i runs over the divisors of n, n/d_i also runs over these divisors, but in reverse order. ▲

To illustrate the theorem and its proof, take $n = 12$. Then

$$\varphi(1) + \varphi(2) + \varphi(3) + \varphi(4) + \varphi(6) + \varphi(12) = 1 + 1 + 2 + 2 + 2 + 4 = 12,$$

$$C(1) = \{1, 5, 7, 11\}, \qquad C(2) = \{2, 10\}, \qquad C(3) = \{3, 9\},$$

$$C(4) = \{4, 8\}, \qquad C(6) = \{6\}, \qquad C(12) = \{12\}.$$

PROBLEMS

1. Prove with the help of Theorem 3-8 that if $(a, b) = d$, then

$$\varphi(ab) = \frac{d\varphi(a)\varphi(b)}{\varphi(d)}.$$

2. Show that if $n > 1$, then the sum of the positive integers less than n and prime to it is

$$\frac{n\varphi(n)}{2}.$$

[*Hint:* If m satisfies the conditions, so does $n - m$.]

3. Show that if $d|n$, then $\varphi(d)|\varphi(n)$.

4. Let n be positive. Show that any solution of the equation

$$\varphi(x) = 4n + 2$$

is of one of the forms p^α or $2p^\alpha$, where p is a prime of the form $4s - 1$. Deduce that there is no solution of the equation $\varphi(x) = 14$. [*Hint:* Use the factorization of $\varphi(x)$ as given in Theorem 3-8.]

*5. Let $f(x)$ be a polynomial with integral coefficients, and let $\psi(n)$ denote the number of values

$$f(0), f(1), \ldots, f(n - 1)$$

which are prime to n.

(a) Show that ψ is multiplicative:

$$\psi(mn) = \psi(m) \cdot \psi(n) \qquad \text{if} \qquad (m, n) = 1.$$

(b) Show that

$$\psi(p^\alpha) = p^{\alpha-1}(p - b_p),$$

where b_p is the number of integers $f(0), f(1), \ldots, f(p - 1)$ which are divisible by the prime p.

6. How many fractions r/s are there satisfying the conditions

$$(r, s) = 1, \qquad 0 \le r < s \le n?$$

7. (a) Use Theorem 3-6 to show that if $p \nmid a$, then the congruence $ax \equiv b \pmod{p}$ is solvable. (Consider separately the cases $p|b$ and $p \nmid b$.) (b) What connection does (a) have with the multiplication table (mod 5) of the preceding section?

8. It follows from Theorem 3-4 that if $a \equiv b \pmod{m}$, then $a^n \equiv b^n \pmod{m}$. Is it always true that if $u \equiv v \pmod{m}$, then $a^u \equiv a^v \pmod{m}$? Construct tables of the smallest positive residues of $a, a^2, a^3, \ldots \pmod{m}$ for $m = 7, 12, 18$, with a running over a complete or reduced residue system (mod m). Make conjectures concerning the periodicity, length of period, etc., of the powers of a fixed number (mod m).

*9. A complex number ξ is said to be an nth *root of unity* if $\xi^n = 1$, and a *primitive* nth root of unity if, in addition, $\xi^m \neq 1$ for $0 < m < n$.

(a) Show that the powers of ξ, a primitive nth root of unity, form a periodic sequence, of period n.

(b) If $(m, n) = d$, show that $\eta = \xi^m$ is an (n/d)th root of unity.

(c) Show that in fact η is a primitive (n/d)th root of unity. [*Hint:* Suppose that n is a primitive rth root, and apply Theorem 1–1.]

(d) Supposing that there is at least one, how many primitive nth roots of unity are there? (Remember that the equation $x^n - 1 = 0$ has only n complex roots.)

*10. Show that for each n the equation $\varphi(x) = n$ has only finitely many solutions. [*Hint:* Show that $\varphi(p^\alpha)$ can be made arbitrarily large by making p^α sufficiently large.]

3–5 Linear congruences. Because of the analogy between congruences and equations, it is natural to ask about the solution of congruences involving one or more (integral) unknowns. In the case of an algebraic congruence $f(x) \equiv 0 \pmod{m}$, where $f(x)$ is a polynomial in x with integral coefficients, we see by Theorem 3–4 that if $x = a$ is a solution, so is every element of the residue class containing a. For this reason it is customary, for such congruences, to list only the solutions between 0 and $m - 1$, inclusive, with the understanding that any x congruent to one of those listed is also a solution. Similarly, when the number of roots of a certain congruence is mentioned, it is actually the number of residue classes that is meant. Attention must be given, however, to the modulus with respect to which the solutions are counted since, for example, the arithmetic progression $\ldots, -3, -1, 1, 3, 5, \ldots$ constitutes a single residue class (mod 2), but two residue classes (mod 4): the elements $\ldots, -3, 1, 5, \ldots$ make up the class of integers $\equiv 1 \pmod{4}$, and the remaining ones the class of integers $\equiv 3 \pmod{4}$.

When we list residue classes as solutions of a congruence, we are, in effect, doing exactly the same thing as when we solve an equation. The solution of the equation $5x = 3$ is given by $x = 3/5$; in other words, the solution of an equation is described by another equation, in which x occurs as one side of the equation and the other does not involve x at all. Similarly, a solution of the congruence $5x \equiv 3 \pmod{7}$ is given by $x \equiv 2 \pmod{7}$; thus the solution of a congruence is described by another congruence, where x appears alone on one side and the other side is free of x.

With these remarks in mind, let us proceed to the details. The simplest case to treat is the linear congruence in one unknown; that is, the congruence

$$ax \equiv b \pmod{m}.$$

As we have already noticed, this is equivalent to the linear Diophantine equation

$$ax - my = b.$$

By Theorem 2–6 this equation can be solved if and only if $(a, m)|b$, and if it is solvable and if x_0, y_0 is a solution, then a general solution is

$$x \equiv x_0 \left(\text{mod } \frac{m}{d}\right), \qquad y \equiv y_0 \left(\text{mod } \frac{a}{d}\right),$$

where $d = (a, m)$. In particular, x is unique modulo m/d. Among the numbers x satisfying the first of these congruences, the numbers

$$x_0, x_0 + \frac{m}{d}, x_0 + \frac{2m}{d}, \ldots, x_0 + \frac{(d-1)m}{d}$$

are incongruent (mod m), whereas all other such numbers x are congruent (mod m) to one of these. Hence we have the following theorem:

THEOREM 3–10. *A necessary and sufficient condition that the congruence*

$$ax \equiv b \pmod{m}$$

be solvable is that $(a, m)|b$. *If this is the case, there are exactly* (a, m) *solutions* (mod m).

While Theorem 3–10 asserts the existence of a solution under appropriate circumstances and predicts the number of such solutions, it says nothing about the process of finding them. If no solution can be found by inspection, then the simplest procedure is to convert the congruence to an equation and solve by the method given at the beginning of Section 2–4.

Consider, for example, the congruence

$$34x \equiv 60 \pmod{98}.$$

Since $(34, 98) = 2$ and $2|60$, there are just two solutions, to be found from

$$17x \equiv 30 \pmod{49}.$$

This is equivalent to $17x - 49y = 30$, and we obtain

$$x = \frac{49y + 30}{17} = 3y + 2 - \frac{2y + 4}{17}, \qquad t = \frac{2y + 4}{17},$$

$$y = \frac{17t - 4}{2} = 8t - 2 + \frac{t}{2}, \qquad z = \frac{t}{2},$$

$$t = 2z.$$

Take $z = 0$; then $t = 0$, $y = -2$, $x = -4$. Hence

$$x \equiv -4 \pmod{49},$$

and the two solutions of the original congruence are

$$x \equiv -4, 45 \pmod{98}.$$

Theorem 3–10 provides the answer to a question alluded to earlier, namely, When is division possible in arithmetic (mod m)? We see now that an integer a has a reciprocal (mod m)—that is, the congruence $ax \equiv 1 \pmod{m}$ is solvable—if and only if $(a, m) = 1$, and correspondingly that the "reduced fraction" b/a makes sense modulo m if and only if $(a, m) = 1$. If m is a prime p, then division by any number not in the residue class of 0 is possible, but for composite modulus this is not the case. Thus

$$\tfrac{3}{4} \equiv 6 \pmod{7} \qquad \text{since} \qquad 3 \equiv 4 \cdot 6 \pmod{7},$$

and

$$\tfrac{3}{5} \equiv 3 \pmod{6} \qquad \text{since} \qquad 3 \equiv 3 \cdot 5 \pmod{6},$$

while $\frac{3}{4}$ has no meaning (mod 6), since $(4, 6)\nmid 3$ and there is no solution of $3 \equiv 4x \pmod{6}$.

Actually, it is customary not to use the fractional notation, but to refer instead to the solution of a linear congruence. This is rather like referring to "the solution of the equation $4x = 3$" instead of "the rational number $\frac{3}{4}$"; neither is logically superior to the other. In the modular case, however, all of the infinitely many "fractions" a/b which make sense (mod m) are congruent to one or another of the elements of the finite set $0, 1, \ldots, m - 1$, and there are obvious advantages in dealing with the smaller set.

The solution of a linear congruence in more than one unknown can be effected by the successive solution of a (usually large) number of congruences in a single unknown. Consider the congruence

$$a_1x_1 + a_2x_2 + \cdots + a_nx_n \equiv c \pmod{m}.$$

The obviously necessary condition for solvability, that $(a_1, \ldots, a_n, m)$ should divide c, is also sufficient, just as in the former case. For, assuming it satisfied, we can divide through by $(a_1, \ldots, a_n, m)$ to get

$$a'_1x_1 + \cdots + a'_nx_n \equiv c' \pmod{m'},$$

where now $(a'_1, \ldots, a'_n, m') = 1$. If $(a'_1, \ldots, a'_{n-1}, m') = d'$, we must have

$$a'_nx_n \equiv c' \pmod{d'};$$

since $(a'_n, d') = 1$, this congruence has just one solution (mod d'), and

m'/d' solutions (mod m'). Substituting each of these in turn yields m'/d' congruences in $n - 1$ unknowns, and the process can be repeated.

As an example, consider the congruence

$$2x + 7y \equiv 5 \pmod{12}.$$

Here $(2, 7, 12) = 1$. Since $(2, 12) = 2$, we must have

$$7y \equiv 5 \pmod{2},$$

which clearly gives $y \equiv 1 \pmod{2}$, or $y \equiv 1, 3, 5, 7, 9, 11 \pmod{12}$. These yield

$$2x \equiv 10, 8, 6, 4, 2, 0 \pmod{12},$$

respectively, or

$$x \equiv 5, 4, 3, 2, 1, 0 \pmod{6}.$$

Thus the solutions (mod 12) are

$$x, y \equiv 5, 1;\ 11, 1;\ 4, 3;\ 10, 3;\ 3, 5;\ 9, 5;\ 2, 7;\ 8, 7;\ 1, 9;\ 7, 9;\ 0, 11;\ 6, 11.$$

The general situation is described in the following theorem, which is easily proved by induction on the number of unknowns.

THEOREM 3–11. The congruence

$$a_1x_1 + \cdots + a_nx_n \equiv c \pmod{m}$$

has just dm^{n-1} or no solutions (mod m), depending on whether $d|c$ or $d\nmid c$, where $d = (a_1, \ldots, a_n, m)$.

Turning now to the simultaneous solution of a system of linear congruences, we consider the system

$$\alpha_1 x \equiv \beta_1 \pmod{m_1}, \ldots, \alpha_n x \equiv \beta_n \pmod{m_n},$$
$$\alpha_i \text{ and } \beta_i \text{ integers.}$$

Clearly, no x satisfies all these congruences unless each can be solved separately. Assuming that this is so, we can suppose that each has already been solved for x, so that we have one or more systems of the form

$$x \equiv c_1 \pmod{m_1}, \ldots, x \equiv c_n \pmod{m_n}.$$

It is obvious that such a system of n congruences will have no solution unless every pair taken from among them has a solution. From the first of the congruences

$$x \equiv c_i \pmod{m_i}, \qquad x \equiv c_j \pmod{m_j},$$

we get $x = c_i + m_i y$; substituting in the second yields

$$m_i y \equiv c_j - c_i \pmod{m_j},$$

and consequently it must be true that

$$(m_i, m_j) | (c_i - c_j).$$

It can be shown that this condition, which is necessary for solvability, is also sufficient: *if* $(m_i, m_j) | (c_i - c_j)$ *for every two indices* i *and* j *between* 1 *and* n *inclusive, then the system* $x \equiv c_i \pmod{m_i}$, $i = 1, \ldots, n$, *is solvable, and the solution is unique modulo the* LCM *of* $m_1, \ldots, m_n$. We shall not carry through the proof of this general statement, but content ourselves with the following special, but extremely important, case:

THEOREM 3–12 (*Chinese Remainder Theorem*). Every system of linear congruences $x \equiv c_1 \pmod{m_1}, \ldots, x \equiv c_n \pmod{m_n}$, in which the moduli are relatively prime in pairs, is solvable, and the solution is unique modulo the product of the moduli.

Proof: The theorem is trivially true if there is only one congruence in the system. Suppose that it is true for every system containing fewer than n congruences, and consider the system $x \equiv c_i \pmod{m_i}$, $i = 1, \ldots, n$, in which $(m_i, m_j) = 1$ for $1 \leq i < j \leq n$. Then by the induction hypothesis we can solve the last $n - 1$ of the congruences simultaneously, and obtain in their place a single congruence with modulus $m_2 \cdots m_n$. Put $m_2 \cdots m_n = M$. Then the system of n congruences is equivalent to the simpler system

$$x \equiv c_1 \pmod{m_1}, \qquad x \equiv C \pmod{M},$$

for suitable C. Repeating the reasoning used above, we find

$$x = c_1 + m_1 y,$$

$$c_1 + m_1 y \equiv C \pmod{M},$$

$$m_1 y \equiv C - c_1 \pmod{M},$$

and since $(m_1, M) = 1$, this last congruence has a unique solution (mod M) by Theorem 3–10. If the solution is $y \equiv C' \pmod{M}$, then we have

$$y = C' + Mz,$$

$$x = c_1 + m_1(C' + Mz) = (c_1 + m_1 C') + m_1 M z,$$

and so x is unique modulo $m_1 M = m_1 \cdots m_n$. ▲

Consider for example the system:

$$x \equiv 1 \pmod 3,$$
$$x \equiv 5 \pmod 8,$$
$$x \equiv 11 \pmod{17}.$$

From the first congruence, $x = 3t + 1$, and from the second,

$$3t \equiv 4 \pmod 8,$$

so that

$$t \equiv 4 \pmod 8,$$
$$t = 8u + 4,$$
$$x = 24u + 13.$$

Using the third of the original congruences, we obtain

$$24u \equiv -2 \pmod{17},$$
$$12u \equiv -1 \pmod{17},$$
$$u \equiv 7 \pmod{17},$$
$$u = 17v + 7,$$
$$x = 24 \cdot 17v + 7 \cdot 24 + 13,$$
$$x \equiv 181 \pmod{3 \cdot 8 \cdot 17}.$$

Problems

1. Find the solutions of each of the following congruences, and check your conclusion against that of Theorem 3–10:

(a) $33x \equiv 21 \pmod{105}$,
(b) $33x \equiv 22 \pmod{105}$,
(c) $15x \equiv 30 \pmod{105}$.

2. Solve the congruence $6x + 15y \equiv 9 \pmod{18}$.

3. Solve simultaneously:

$$x \equiv 1 \pmod 2,$$
$$x \equiv 2 \pmod 3,$$
$$x \equiv 3 \pmod 5,$$
$$x \equiv 5 \pmod 7.$$

4. Suppose that the system of congruences,

$$\begin{cases} x \equiv a_1 \pmod{m_1}, \\ \quad\vdots \\ x \equiv a_n \pmod{m_n}, \end{cases} \tag{*}$$

is to be solved, where $m_1m_2\cdots m_n = M$ and $(m_i, m_j) = 1$ for all i and j with $i \neq j$. For simplicity we write $x \equiv \{a_1, \ldots, a_n\} \pmod{\{m_1, \ldots, m_n\}}$ as an abbreviation for (*). In the following, we present a method for writing the solution of (*) as a combination of solutions of simpler systems.

(a) Solve the system $x \equiv \{0, 0, \ldots, 0\} \pmod{\{m_2, \ldots, m_n\}}$ and thus replace the system $x \equiv \{1, 0, \ldots, 0\} \pmod{\{m_1, m_2, \ldots, m_n\}}$ by a system of two congruences. Replace these two congruences in turn by a single congruence $\pmod{m_1}$, with a new unknown.

(b) Proceeding as in part (a), replace each of the following systems by a single congruence:

$$\begin{aligned} x_1 &\equiv \{1, 0, 0, \ldots, 0\} \pmod{\{m_1, \ldots, m_n\}}; \\ x_2 &\equiv \{0, 1, 0, \ldots, 0\} \pmod{\{m_1, \ldots, m_n\}}; \\ &\vdots \\ x_n &\equiv \{0, 0, \ldots, 0, 1\} \pmod{\{m_1, \ldots, m_n\}}. \end{aligned}$$

(c) Let the separate congruences obtained in part (b) have solutions $x_1 \equiv e_1 \pmod{M}, \ldots, x_n \equiv e_n \pmod{M}$, respectively. Show that

$$x \equiv a_1e_1\frac{M}{m_1} + a_2e_2\frac{M}{m_2} + \cdots + a_ne_n\frac{M}{m_n} \pmod{M}$$

is the solution of the original system (*). Note that once the numbers $e_1, \ldots, e_n$ have been computed, the additional work required by a change in the constants $a_1, \ldots, a_n$ is almost nil.

5. Apply the procedure of Problem 4 to solve the following systems:

$$\text{(a)}\quad \begin{aligned} x &\equiv 10 \pmod{27}, \\ x &\equiv 2 \pmod{25}, \\ x &\equiv 3 \pmod{8}; \end{aligned} \qquad \text{(b)}\quad \begin{aligned} x &\equiv 8 \pmod{27}, \\ x &\equiv 21 \pmod{25}, \\ x &\equiv 1 \pmod{8}. \end{aligned}$$

6. Solve the following systems:

$$\text{(a)}\ \begin{aligned} x &\equiv 1 \pmod{8}, \\ x &\equiv 43 \pmod{81}, \end{aligned} \qquad \text{(b)}\ \begin{aligned} x &\equiv 5 \pmod{8}, \\ x &\equiv 73 \pmod{81}, \end{aligned} \qquad \text{(c)}\ \begin{aligned} x &\equiv 4 \pmod{8}, \\ x &\equiv 19 \pmod{81}. \end{aligned}$$

7. A famous theorem of P. L. Dirichlet asserts that if k and l are relatively prime, then there are infinitely many primes of the form $kx + l$. The proof is rather difficult. Prove the much weaker statement that if $(k, l) = 1$ and $n \neq 0$, there is an x such that $(kx + l, n) = 1$. [*Hint:* It must be shown that x can be chosen such that every prime p dividing n does not divide $kx + l$. Treat separately the cases $p \mid l$ and $p \nmid l$.]

8. Show that Dirichlet's theorem implies, and is implied by, the following assertion: if $(k, l) = 1$, then there is at least one prime of the form $kx + l$.

3–6 Polynomial congruences. As is well known, the equality sign is used between polynomials in two essentially different ways. In the equation

$$(x + a)^2 = x^2 + 2ax + a^2,$$

for example, it means that the left- and right-hand sides are identical polynomials, i.e., that the coefficients of x^2, and of x, are equal, as are the constant terms. In the equation $x^2 - 2 = 0$, it means that the square of the number x is equal to 2, and this may be true or false for particular x. If we temporarily refer to the first as algebraic equality, and to the second as numerical equality, there is the following connection between them: if two polynomials are algebraically equal, they are also numerically equal for every value of x, and if two polynomials are numerically equal for every value of x (or even for infinitely many values of x), then they are algebraically equal.

The congruence symbol is also used in two different ways, to relate polynomials. When $f(x)$ and $f_1(x)$ are polynomials, we write

$$f(x) \equiv f_1(x) \pmod{m} \tag{1}$$

if the coefficients of each power of x in f and f_1 are congruent (mod m). For example,

$$(x + a)^2 \equiv x^2 + a^2 \pmod{2},$$

and

$$x(x - 1) \equiv (x - 3)(x + 2) \pmod{6}. \tag{2}$$

This meaning of the congruence symbol is usually intended when there is no reference to numerical values of x, or to roots or solutions of the congruence. The other meaning of the symbol is that x is a number for which the numerical values $f(x)$ and $f_1(x)$ are congruent (mod m).

It should be noted that the connection indicated above between the two meanings of equality does not extend to congruences; that is, (1) is not equivalent to

$$f(x) \equiv f_1(x) \pmod{m} \qquad \text{for all } x,$$

since, for example, $x^3 \equiv x \pmod{3}$ for all x, whereas obviously x^3 and x are not "algebraically" congruent (mod 3).

There are other ways as well in which polynomial congruences behave differently from polynomial equations. It is a theorem (though possibly not one familiar to the reader) that every polynomial with integral coefficients factors in a unique way into irreducible polynomials with integral coefficients. The congruence (2) above shows that this is no longer true for polynomials whose coefficients are integers (mod 6), since the two factorizations given for $x^2 - x$ are genuinely different, while the linear factors are clearly incapable of further factorization, and so are irreducible. We also see from (2) that there is no general analog of the familiar theorem from algebra that the number of roots of a polynomial equation is equal to the degree of the polynomial, since the quadratic congruence $x^2 - x \equiv 0 \pmod{6}$ has the four solutions $x \equiv 0, 1, 3, 4 \pmod{6}$. If there were fewer solutions than the degree would indicate, there might be some hope of finding further ones by considering larger number systems, in much the same way that the equation $x^2 + 1 = 0$, which is not solvable in integers, or even in rational numbers or in real numbers, becomes solvable when complex numbers are allowed. But there is not much to be done when there are too many solutions.

The reverse situation, in which there are too few solutions, also occurs, of course. The congruence $x^2 + 1 \equiv 0 \pmod{m}$ has two solutions when $m = 5$, namely $x \equiv \pm 2 \pmod{5}$, but it has none for $m = 7$. These examples show also that the strong distinction between real and complex numbers sometimes disappears in this "modular" arithmetic, since -1 already has a square root (mod 5)!

It turns out that some degree of order returns if we restrict attention to congruences with prime moduli: polynomials have unique factorization, they have no more roots than the degree would indicate, etc. For this reason we shall frequently consider theorems valid only for this particular case, although some, such as the following one, are true in general.

THEOREM 3–13 (*Factor theorem*). If a is a root of the congruence

$$f(x) \equiv 0 \pmod{m},$$

then there is a polynomial $g(x)$ such that

$$f(x) \equiv (x - a)g(x) \pmod{m},$$

and conversely.

Proof: Suppose first that a is a root of $f(x) \equiv 0 \pmod{m}$, and apply the ordinary long-division algorithm learned in algebra to divide $f(x)$ by $x - a$. Since the leading coefficient in the divisor is 1, no fractions will ever be encountered in the process. Since the divisor is of degree 1, the process can be continued until the remainder is a constant, say r, and we obtain an identity of the sort

$$f(x) = (x - a)g(x) + r.$$

The polynomials on the left and right are algebraically equal, and therefore they are algebraically congruent modulo m:

$$f(x) \equiv (x - a)g(x) + r \pmod{m}.$$

Setting $x \equiv a \pmod{m}$, we have

$$f(a) \equiv 0 \equiv 0 \cdot g(a) + r \pmod{m},$$

whence $r \equiv 0 \pmod{m}$, and

$$f(x) \equiv (x - a)g(x) \pmod{m},$$

as asserted. If conversely this last congruence holds, then clearly $f(a) \equiv 0 \pmod{m}$. ▲

THEOREM 3-14 (*Lagrange's theorem*). The congruence

$$f(x) \equiv 0 \pmod{p}$$

in which

$$f(x) = a_0x^n + \cdots + a_n, \qquad a_0 \not\equiv 0 \pmod{p},$$

has at most n roots.

Proof: For $n = 1$, the assertion follows from Theorem 3-10. Assume that every congruence of degree $n - 1$ has at most $n - 1$ solutions, and that a is a root of the original congruence. Then

$$f(x) \equiv (x - a)q(x) \pmod{p},$$

where $q(x)$ is of exact degree $n - 1$. By the induction hypothesis, $q(x)$ therefore has at most $n - 1$ zeros, say $c_1, \ldots, c_r$, where $r \leq n - 1$. Thus, if c is any number such that $f(c) \equiv 0 \pmod{p}$, then

$$(c - a)q(c) \equiv 0 \pmod{p},$$

so that either

$$c \equiv a \pmod{p}$$

or

$$q(c) \equiv 0 \pmod{p}.$$

In the latter case, $c = c_i$ for some i, $1 \leq i \leq r$. In other words, the original congruence has at most $r + 1 \leq n$ roots. The theorem now follows by the induction principle. ▲

PROBLEMS

1. Let $f(x)$ be a polynomial of degree n, with integral coefficients. Show that if $n + 1$ consecutive values of $f(x)$ are divisible by a fixed prime p, then $p|f(x)$ for every integral x. Cf. Problem 1, Section 3-2.

2. Find all solutions of $x^{12} \equiv 1 \pmod{13}$. [The computation of high powers is best accomplished by using the binary expansion of the exponent, e.g.,

$$\begin{aligned} 2^2 &\equiv 4, \\ 2^4 &\equiv 4^2 \equiv 3, \\ 2^8 &\equiv 3^2 \equiv 9, \\ 2^{12} &\equiv 2^{8+4} \equiv 27 \equiv 1 \pmod{13}.] \end{aligned}$$

3. Using only the number of solutions in the preceding problem, and ignoring any information about the behavior of individual solutions, show that if $d|12$ and $d < 12$, then the congruence $x^d \equiv 1 \pmod{13}$ has exactly d solutions. [*Hint:* Factor $x^{12} - 1$ as $(x^d - 1)q(x)$ and apply Lagrange's theorem.]

The following seven problems sketch a proof of the unique factorization theorem for polynomials (mod p). If $a(x)$ is a polynomial, we shall mean by $\deg_p a(x)$ the degree of the highest-degree term in which the coefficient is not divisible by p. In particular, if a is a constant not divisible by p, then $\deg_p a = 0$. If $p|a$, the symbol $\deg_p a$ is not defined.

4. Let $a(x)$ and $b(x)$ be nonzero polynomials with $\deg_p a(x) = n \geq \deg_p b = m$. Show that for a suitable constant c_0, either $\deg_p \big(a(x) - c_0x^{n-m}b(x)\big) < n$ or $a(x) - c_0x^{n-m}b(x) \equiv 0 \pmod{p}$.

5. Prove by induction that if $a(x)$ and $b(x)$ are nonzero polynomials (mod p), then there are $q(x)$ and $r(x)$ such that $a(x) \equiv b(x)q(x) + r(x) \pmod{p}$ and either $r(x) \equiv 0 \pmod{p}$ or $\deg_p r(x) < \deg_p b(x)$. [*Hint:* Put $r_1(x) = a(x) - c_0x^{n-m}b(x)$; if $\deg_p r_1(x) \geq \deg_p b(x)$, the procedure of Problem 4 can be repeated.]

6. Let $a(x)$ and $b(x)$ be nonzero polynomials (mod p), and let $d(x)$ be any polynomial of minimum degree such that for some $g(x)$ and $h(x)$,

$$a(x)g(x) + b(x)h(x) \equiv d(x) \pmod{p}.$$

(a) Show that any two polynomials $d(x)$ and $d_0(x)$ with these properties differ only by a constant factor. [*Hint:* By an appropriate choice of $g(x)$, $g_0(x)$, $h(x)$ and $h_0(x)$, the leading coefficients of both $d(x)$ and $d_0(x)$ can be made 1. Consider $d(x) - d_0(x)$.]

(b) Show that if $d_1(x)|_p a(x)$ [that is, if $a(x) \equiv d_1(x)q(x) \pmod{p}$ for suitable $q(x)$] and $d_1(x)|_p b(x)$, then $d_1(x)|_p d(x)$.

(c) Show that $d(x)|_p a(x)$ and $d(x)|_p b(x)$. (See Problem 12, Section 2–2.) We write $\big(a(x),b(x)\big)_p = d(x)$.

7. Show that every polynomial $a(x)$ with $\deg_p a(x) \geq 1$ can be represented as a product of one or more polynomials irreducible (mod p).

8. Show that if $a(x)|_p b(x)c(x)$ and $\big(a(x), b(x)\big)_p = 1$, then $a(x)|_p c(x)$.

9. Show that if $P(x)$, $P_1(x)$, ..., $P_n(x)$ are polynomials irreducible (mod p), and $P(x)|_p \prod_{i=1}^r P_i(x)$, then for at least one i, $P(x)|_p P_i(x)$.

10. Show that the representation of a polynomial $a(x)$ with $\deg_p a(x) \geq 1$ as a product of polynomials irreducible (mod p) is unique, except for the order of

factors and the presence of constant factors. [Note that $b(x)|_p c(x)$ and $c(x)|_p b(x)$ together imply only that $b(x) \equiv A \cdot c(x) \pmod p$ for some constant A.]

11. Find an example of polynomials $a(x)$ and $b(x)$ which are relatively prime (mod p) but not (mod q), for suitable primes p and q.

3–7 Quadratic congruences with prime modulus. The general problem of higher-degree congruences is too difficult for further development here, but we can obtain some information in an elementary way in the special case of quadratic congruences. Consider the congruence

$$ax^2 + bx + c \equiv 0 \pmod p, \qquad p \nmid a, \tag{3}$$

where it is now supposed that p is an odd prime. (The case $p = 2$ is of no particular interest, since the only distinct quadratic polynomials are then x^2, $x^2 + 1$, $x^2 + x$ and $x^2 + x + 1$, and the solutions of the corresponding congruences can be given explicitly.) Since $p \nmid 4a$, the congruence (3) is equivalent to

$$4a^2x^2 + 4abx + 4ac \equiv 0 \pmod p,$$

and hence to

$$(2ax + b)^2 \equiv b^2 - 4ac \pmod p.$$

Let $b^2 - 4ac = d$. If the congruence $u^2 \equiv d \pmod p$ is not solvable, then neither is (3). On the other hand, if u_1 is a number such that $u_1^2 \equiv d \pmod p$, then the integer x_1 such that $2ax_1 + b \equiv u_1 \pmod p$ is a solution of (3). Conversely, every solution of (3) is related to a solution of $u^2 \equiv d \pmod p$ by such a linear congruence $2ax + b \equiv u \pmod p$. Since this linear congruence has exactly one solution x for each u, we see that there is a one-to-one correspondence between the solutions of (3) and those of $u^2 \equiv d \pmod p$, and we may as well restrict our attention to the latter.

Changing the notation slightly, consider the congruence

$$x^2 \equiv a \pmod p. \tag{4}$$

If this congruence is solvable, it would be reasonable to say that a is a square, modulo p, but for historical reasons the customary phrase is "quadratic residue": a is a quadratic residue of p if (4) is solvable; otherwise a is a quadratic nonresidue. (Analogous definitions hold for nth-power residues and nonresidues.) We shall now develop a criterion for deciding whether a is a quadratic residue of p:

THEOREM 3–15 (*Euler's criterion*). A necessary and sufficient condition that a be a quadratic residue of the odd prime p is that the congruence

$$a^{(p-1)/2} \equiv 1 \pmod p$$

hold.

Proof: Instead of (4), we first consider the congruence

$$bx \equiv a \pmod{p}, \tag{5}$$

in which b is one of the numbers $1, 2, \ldots, p - 1$. This linear congruence is always solvable, since $p \nmid b$, and the solution is unique if we require that it also be one of the numbers $1, 2, \ldots, p - 1$. Let the solution be $x = b'$. For fixed a, the numbers b and b' will be called associates. We must distinguish two cases, depending on whether some b is associated with itself or not.

If for some b, say b_1, we have $b_1 = b_1'$, then (5) becomes $b_1^2 \equiv a \pmod{p}$, and we have a solution of (4). Furthermore, in this case, $(p - b_1)^2 = p^2 - 2pb_1 + b_1^2 \equiv b_1^2 \equiv a \pmod{p}$, and since $b_1 \not\equiv p - b_1$, we obtain two distinct solutions of (4). By Lagrange's theorem there are no others, so that for all b different from b_1 and $p - b_1$, we find that b is different from its associate b'. Thus if a is a quadratic residue of p, the integers $1, \ldots, p - 1$ can be grouped into $(p - 3)/2$ pairs of distinct associates, the product of the elements of each pair being congruent to $a \pmod{p}$, together with the two numbers b_1 and $p - b_1$. Hence

$$(p - 1)! = \prod_{k=1}^{p-1} k \equiv a^{(p-3)/2} \cdot b_1(p - b_1) \equiv -a^{(p-1)/2} \pmod{p}. \tag{6}$$

On the other hand, if a is a quadratic nonresidue of p, the numbers $1, 2, \ldots, p - 1$ can be grouped into $(p - 1)/2$ pairs of distinct associates, and

$$(p - 1)! = \prod_{k=1}^{p-1} k \equiv a^{(p-1)/2} \pmod{p}. \tag{7}$$

In order to give a uniform statement of (6) and (7), we define the *Legendre symbol* (a/p) [also frequently written $\left(\frac{a}{p}\right)$ or $(a|p)$] to mean 1 if a is a quadratic residue of p, and -1 if a is a quadratic nonresidue of p. [Here a is called the "first entry," and p the "second entry." Note that (a/p) is not defined if $p|a$.] Then (6) and (7) become

$$(p - 1)! \equiv -(a/p)a^{(p-1)/2} \pmod{p}. \tag{8}$$

Taking $a = 1$, and noting that the congruence $x^2 \equiv 1 \pmod{p}$ has the solution $x = 1$, so that $(1/p) = 1$, we have $(p - 1)! \equiv -1 \pmod{p}$. Substituting in (8) yields

$$(a/p)a^{(p-1)/2} \equiv 1 \pmod{p},$$

or since $(a/p) = \pm 1$,

$$(a/p) \equiv a^{(p-1)/2} \pmod{p}. \blacktriangle$$

In the course of the proof, we also obtained the following theorem:

THEOREM 3–16 (*Wilson's Theorem*). If p is prime, then $(p-1)! \equiv -1 \pmod{p}$.

(Strictly speaking, we have proved this only for odd p; but the proof is trivial for $p = 2$.)

PROBLEMS

1. Show that if $p \equiv 1 \pmod 4$, then $(a/p) = (-a/p)$.
2. Evaluate the Legendre symbols $(5/17)$, $(6/31)$, and $(8/11)$.
3. For what primes p is the congruence $x^2 \equiv -1 \pmod{p}$ solvable?
4. Show that if $p \nmid a$ and $p \nmid b$, then

(a) $(a^2/p) = 1$,
(b) $(a/p) = (b/p)$ if $a \equiv b \pmod{p}$,
(c) $(ab/p) = (a/p)(b/p)$.

5. Solve the following congruences, or show them to be unsolvable:

(a) $3x^2 - 5x + 7 \equiv 0 \pmod{13}$,
(b) $5x^2 - 6x + 2 \equiv 0 \pmod{13}$,
(c) $x^2 + 7x + 10 \equiv 0 \pmod{11}$.

6. Use the fact that $j \equiv -(p-j) \pmod{p}$ to pair off the factors in $(p-1)!$, and thus obtain from Wilson's theorem a solution of $x^2 \equiv -1 \pmod{p}$ when $p \equiv 1 \pmod 4$.

CHAPTER 4

THE POWERS OF AN INTEGER, MODULO m

4–1 The order of an integer (mod m). The sequence of powers of a fixed positive integer a is a special case of the more general geometric progressions studied in algebra. These successive powers are distinct integers if $a > 1$, and they increase quite rapidly. We shall now study the sequence which results when the powers are all reduced to their least positive remainders (mod m), where m is an integer relatively prime to a. Here again, as in the preceding chapter, there are problems whose solutions for composite modulus are too complicated for inclusion in this text, and these will be discussed only for prime modulus. The reader should take care to be aware of this restriction whenever it is present.

We begin with a specific case, the sequence of powers of 2, reduced (mod 17). The following congruences hold, the modulus 17 being understood throughout:

$$\begin{aligned}
2^0 &\equiv 1,\\
2^1 &\equiv 2,\\
2^2 &\equiv 4,\\
2^3 &\equiv 8,\\
2^4 &\equiv 16,\\
2^5 &\equiv 2 \cdot 16 = 32 \equiv 15,\\
2^6 &\equiv 2 \cdot 15 = 30 \equiv 13,\\
2^7 &\equiv 2 \cdot 13 = 26 \equiv 9,\\
2^8 &\equiv 2 \cdot 9 = 18 \equiv 1,\\
2^9 &\equiv 2 \cdot 1 = 2,\\
2^{10} &\equiv 4,\\
&\vdots
\end{aligned}$$

We see that there is no point in continuing further, because the sequence is already repeating itself; since $2^8 \equiv 1 \pmod{17}$, we have $2^{8+j} \equiv 2^8 \cdot 2^j \equiv 2^j \pmod{17}$, and hence any two powers of 2 whose exponents differ by 8 (or a multiple of 8) are congruent to each other (mod 17). In other words, the sequence is periodic from the beginning, with period 8. The length of the period is the smallest positive exponent n such that $2^n \equiv 1 \pmod{17}$.

What has happened here is entirely typical: if $(a, m) = 1$, the least positive residues (mod m) of $a^0, a^1, \ldots$ always form a periodic sequence, and this sequence is always periodic from the beginning. To see this, note first that while there are infinitely many powers of a, there are only the m integers $0, 1, \ldots, m - 1$ for them to be congruent to, and hence some two powers of a must be congruent to each other. Suppose that $a^r \equiv a^s \pmod{m}$, where $r > s$. Then $a^s(a^{r-s} - 1) \equiv 0 \pmod{m}$, and since $(a^s, m) = (a, m) = 1$, we must have $a^{r-s} \equiv 1 \pmod{m}$. But then $a^{r-s+1} \equiv a$, $a^{r-s+2} \equiv a^2$, etc., so the sequence is surely periodic.

Moreover, 1, which is the first element of the sequence, is also the first number to repeat. For suppose the opposite: that the second occurrence of 1 is at the power a^n, and that for some r and s with $0 < s < r < n$, we have $a^r \equiv a^s \pmod{m}$. Then, just as before, we can deduce that $a^{r-s} \equiv 1 \pmod{m}$, which contradicts the definition of n, since $0 < r - s < n$.

The most obvious problem remaining, then, is that of determining the length of the period. This length cannot be predicted in general, although for specific a and m it can, of course, be found by computing the sequence. We can, however, get some useful information about the period length, the simplest fact being that it is always less than m; for if it were not, the m numbers $a^0, a^1, \ldots, a^{m-1}$ would be distinct and different from $0 \pmod{m}$, which is clearly impossible.

We call the length of the period the *order of a* (mod m), or the *exponent to which a belongs* (mod m), and we write $\text{ord}_m\, a$; as we have seen, the order can also be defined as the smallest positive integer n such that $a^n \equiv 1 \pmod{m}$. To see what values can be expected for the order of $a \pmod{m}$, consider for example the various sequences of powers (mod 19):

TABLE 4-1

a	a^2	a^3	a^4	a^5	a^6	a^7	a^8	a^9	a^{10}	a^{11}	a^{12}	a^{13}	a^{14}	a^{15}	a^{16}	a^{17}	a^{18}
1																	
2	4	8	16	13	7	14	9	18	17	15	11	3	6	12	5	10	1
3	9	8	5	15	7	2	6	18	16	10	11	14	4	12	17	13	1
4	16	7	9	17	11	6	5	1									
5	6	11	17	9	7	16	4	1									
6	17	7	4	5	11	9	16	1									
7	11	1															
8	7	18	11	12	1												
9	5	7	6	16	11	4	17	1									
10	5	12	6	3	11	15	17	18	9	14	7	13	16	8	4	2	1
11	7	1															
12	11	18	7	8	1												
13	17	12	4	14	11	10	16	18	6	2	7	15	5	8	9	3	1
14	6	8	17	10	7	3	4	18	5	13	11	2	9	12	16	15	1
15	16	12	9	2	11	13	5	18	4	3	7	10	17	8	6	14	1
16	9	11	5	4	7	17	6	1									
17	4	11	16	6	7	5	9	1									
18	1																

Here the orders or period lengths occurring are 1, 2, 3, 6, 9, and 18, i.e., exactly the divisors of 18. Thus we know that the period length always is at most $m - 1$, and that for $m = 19$ all divisors of $m - 1$ occur as period lengths. The latter, however, is not a general phenomenon, for consider the case $m = 10$:

a	a^2	a^3	a^4
1			
3	9	7	1
7	9	3	1
9	1		

The numbers 1, 3, 7, 9 constitute a reduced residue system, and the lengths of the periods of their powers (mod 10) are 1, 2, and 4. Again we have all divisors of a number as period lengths, but this time the number is not $m - 1$. The correct analogy between the two cases is this: when $m = 19$, there are 18 elements in a reduced residue system, and the period lengths all divide 18; when $m = 10$, there are 4 elements in a reduced residue system, and the period lengths all divide 4. In these two cases all divisors actually occur, but this is not always true, as we see by examining the case $m = 12$: $\varphi(12) = 4$, but $5^2 \equiv 7^2 \equiv 11^2 \equiv 1 \pmod{12}$, so that all periods are of length 1 or 2.

We now prove the general theorems bearing on these data.

THEOREM 4–1 (*Fermat's theorem*). If $p \nmid a$, then

$$a^{p-1} \equiv 1 \pmod{p}.$$

Since $\varphi(p) = p - 1$, this is a special case of

THEOREM 4–2 (*Euler's theorem*). If $(a, m) = 1$, then

$$a^{\varphi(m)} \equiv 1 \pmod{m}.$$

Proof: Let $c_1, \ldots, c_{\varphi(m)}$ be a reduced residue system (mod m), and let a be prime to m. Then $ac_1, \ldots, ac_{\varphi(m)}$ is also a reduced residue system (mod m), and therefore

$$\prod_{i=1}^{\varphi(m)} ac_i \equiv \prod_{i=1}^{\varphi(m)} c_i \pmod{m},$$

whence

$$a^{\varphi(m)} \prod_{i=1}^{\varphi(m)} c_i \equiv \prod_{i=1}^{\varphi(m)} c_i \pmod{m}.$$

Since $(m, \Pi c_i) = 1$, this implies that

$$a^{\varphi(m)} \equiv 1 \pmod{m}. \blacktriangle$$

THEOREM 4-3. If $a^u \equiv 1 \pmod{m}$, then $\text{ord}_m\, a|u$.

Proof: Put $\text{ord}_m\, a = t$, and let $u = qt + r$, $0 \leq r < t$. Then

$$a^u = a^{qt+r} = (a^t)^q \cdot a^r \equiv a^r \equiv 1 \pmod{m},$$

and if r were different from zero, there would be a contradiction with the definition of t. ▲

THEOREM 4-4. For every a prime to m, $\text{ord}_m\, a|\varphi(m)$.

Proof: The assertion follows immediately from Theorems 4-1 and 4-3. ▲

It is convenient to prove here a theorem which we shall use in the next section, and which, in a certain sense, generalizes Fermat's theorem. If we consider the polynomial congruence $x^{p-1} \equiv 1 \pmod{p}$, we see from Fermat's theorem that there are exactly $p - 1$ roots, namely $x \equiv 1, \ldots, p - 1 \pmod{p}$, and, by Lagrange's theorem, this is the maximum number of permissible roots. The following theorem introduces other polynomials having the maximum numbers of roots.

THEOREM 4-5. If p is prime and d divides $p - 1$, then there are exactly d roots of the congruence

$$x^d \equiv 1 \pmod{p}.$$

Proof: Since $d|(p - 1)$, we have

$$x^{p-1} - 1 = (x^d - 1)q(x),$$

where $q(x)$ is a polynomial of degree $p - 1 - d$ in x. By Lagrange's theorem, the congruence

$$q(x) \equiv 0 \pmod{p}$$

has at most $p - 1 - d$ solutions. Since $x^{p-1} \equiv 1 \pmod{p}$ has exactly $p - 1$ solutions, $x^d \equiv 1 \pmod{p}$ must have at least

$$p - 1 - (p - 1 - d) = d$$

solutions. Since it can have no more than this number, it must have exactly d solutions. ▲

PROBLEMS

1. Show that if $ab \equiv 1 \pmod{m}$, then

$$\text{ord}_m\, a = \text{ord}_m\, b.$$

2. Show that if p is an odd prime and $\text{ord}_{p^\alpha}\, a = 2t$, then

$$a^t \equiv -1 \pmod{p^\alpha}.$$

Demonstrate that this need not be true if $p = 2$.

3. Show that if p is an odd prime and $a^t \equiv -1 \pmod{p}$, then a belongs to an even exponent $2u \pmod{p}$, and t is an odd multiple of u.

*4. Show that if p is an odd prime and $p|(x^{2^r} + 1)$, then $p \equiv 1 \pmod{2^{r+1}}$. Deduce that there are infinitely many primes congruent to 1 modulo any fixed power of 2.

5. Show that for $a > 1$ and $n > 0$, $n|\varphi(a^n - 1)$.

6. Compute (a) $\text{ord}_{19}\, 12$, (b) $\text{ord}_{31}\, 3$, (c) $\text{ord}_{10}\, 7$.

7. Show that the congruence $f(x) \equiv 0 \pmod{p}$, of degree $m < p$, has m distinct roots if and only if $f(x)|_p(x^p - x)$. [The notation is that of Problem 6(b), Section 3–6.]

8. Find all roots of $x^5 \equiv 1 \pmod{31}$ without computation, using the fact that 2 is a root.

4–2 Integers belonging to a given exponent (mod p).

THEOREM 4–6. If $\text{ord}_m\, a = t$, then $\text{ord}_m\, a^n = t/(n, t)$.

Proof: Let $(n, t) = d$. Then since $a^t \equiv 1 \pmod{m}$, we have

$$(a^t)^{n/d} = (a^n)^{t/d} \equiv 1 \pmod{m},$$

so that if $\text{ord}_m\, a^n = t'$, then

$$t' \,\Big|\, \frac{t}{d}. \tag{1}$$

But from the congruence

$$(a^n)^{t'} \equiv 1 \pmod{m},$$

we have $t|nt'$ by Theorem 4–3, or

$$\frac{t}{d} \,\Big|\, \frac{n}{d}t'.$$

Since

$$\left(\frac{t}{d}, \frac{n}{d}\right) = 1,$$

we obtain

$$\frac{t}{d} \,\Big|\, t'. \tag{2}$$

Combining (1) and (2), we have

$$t' = \frac{t}{d}. \blacktriangle$$

For example, we see from Table 4-1 that

$$\begin{aligned} \operatorname{ord}_{19} 2 &= 18, \\ \operatorname{ord}_{19} 2^2 &= 9, \\ \operatorname{ord}_{19} 2^3 &= 6, \\ \operatorname{ord}_{19} 2^4 &= 9, \\ \operatorname{ord}_{19} 2^5 &= \operatorname{ord}_{19} 13 = 18, \end{aligned}$$

and correspondingly,

$$\frac{18}{(18, 2)} = 9,$$
$$\frac{18}{(18, 3)} = 6,$$
$$\frac{18}{(18, 4)} = 9,$$
$$\frac{18}{(18, 5)} = 18.$$

We also see from Table 4-1 that the integers having order 18 are exactly the powers 2^a of 2 for which $(18, a) = 1$, namely 2^1, $2^5 \equiv 13$, $2^7 \equiv 14$, $2^{11} \equiv 15$, $2^{13} \equiv 3$, and $2^{17} \equiv 10 \pmod{19}$. This is a special case of the next theorem.

THEOREM 4-7. If any integer belongs to $t \pmod p$, then exactly $\varphi(t)$ incongruent numbers belong to $t \pmod p$.

Proof: Assume that $\operatorname{ord}_p a = t$. Then by Theorem 4-4, $t|(p - 1)$, and hence by Theorem 4-5 there are exactly t roots of the congruence $x^t \equiv 1 \pmod p$. But all the numbers $a, a^2, \ldots, a^t$ are roots of this congruence and since they are incongruent (mod p), they are the only roots. By Theorem 4-6, the powers of a which belong to $t \pmod p$ are the numbers a^n with $(n, t) = 1$, $1 \leq n \leq t$, and there are precisely $\varphi(t)$ of these numbers. ▲

THEOREM 4-8. If $t|(p - 1)$, there are $\varphi(t)$ incongruent numbers (mod p) which belong to $t \pmod p$.

Proof: Let d run over the divisors of $p - 1$, and for each such d let $\psi(d)$ be the number of integers among $1, 2, \ldots, p - 1$ of order $d \pmod p$. By Theorem 4-4 and Fermat's theorem, each of the integers $1, 2, \ldots,$ $p - 1$ belongs to exactly one of the d. Hence

$$\sum_{d|(p-1)} \psi(d) = p - 1.$$

But, by Theorem 3–9, we also have

$$\sum_{d|(p-1)} \varphi(d) = p - 1,$$

whence

$$\sum_{d|(p-1)} \psi(d) = \sum_{d|(p-1)} \varphi(d).$$

By Theorem 4–7, the value of $\psi(d)$ is either zero or $\varphi(d)$ for each d, and we deduce from the last equation that $\psi(d) = \varphi(d)$ for each d dividing $p - 1$, since otherwise the first sum would be smaller than the second. ▲

If $\text{ord}_m\, a = \varphi(m)$, then a is said to be a *primitive root* of m. (As noted earlier, for example, the primitive roots of 19 are 2, 3, 10, 13, 14, 15.) The importance of this notion lies in the fact that if g is such a primitive root, then its powers,

$$g, g^2, \ldots, g^{\varphi(m)},$$

are distinct (mod m), and are all relatively prime to m; they therefore constitute a reduced residue system modulo m. Thus we have a convenient way of representing all the elements of a reduced residue system, some implications of which are to be found later in this chapter and in the problems.

It follows immediately from Theorem 4–6 that the other primitive roots of m are those powers g^k for which $(k, \varphi(m)) = 1$. Either from this remark or from Theorem 4–8 we have

THEOREM 4–9. There are exactly $\varphi(p - 1)$ primitive roots of a prime p.

The question of just which moduli have primitive roots is not altogether simple. Without going into details of the proof, we record the answer: the numbers having primitive roots are exactly those of the forms 2, 4, p^α, $2p^\alpha$, where p is any odd prime. We shall use q as a symbol for these numbers throughout the remainder of the present chapter.

The problem of actually finding a primitive root, for large modulus, has not been solved, in the sense that no simple algorithm leads straight to a solution. For given modulus q, it is, of course, a finite problem which can be solved by successively testing the elements of a reduced residue system. A slightly more rapid method is indicated in Problem 3 at the end of the next section, but this is also laborious for large q, particularly if $\varphi(q)$ has many distinct prime divisors.

PROBLEMS

1. Show that if $\text{ord}_p\, a = t$, $\text{ord}_p\, b = u$, and $(t, u) = 1$, then

$$\text{ord}_p\,(ab) = tu.$$

2. Show that if $p \equiv 1 \pmod 4$ and g is a primitive root of p, then so is $-g$. Show by a numerical example that this need not be the case if $p \equiv 3 \pmod 4$.

3. Show that if p is of the form $2^m + 1$ and $(a/p) = -1$, then a is a primitive root of p. [*Hint:* What are the conceivable orders of a?]

4. Show that if p is an odd prime and $\operatorname{ord}_p a = t > 1$, then

$$\sum_{k=1}^{t-1} a^k \equiv -1 \pmod p.$$

5. Show that if q has primitive roots, there are $\varphi(\varphi(q))$ of them, and their product is congruent to $1 \pmod q$ if $q > 6$. [*Hint:* Represent all primitive roots in terms of a single one.]

6. Find all primitive roots of 25.

7. Find a primitive root of 23 and then, using Theorem 4-6, all primitive roots of 23.

4-3 Indices. Let q be a number having primitive roots and let g be one of them. Then the numbers $g, g^2, \ldots, g^{\varphi(q)}$ are distinct $\pmod q$, and they are all prime to q; therefore they constitute a reduced residue system $\pmod q$. The relation between a number a and the exponent of a power of g which is congruent to $a \pmod q$ is very similar to the relation between an ordinary positive real number x and its logarithm. This exponent is called an *index* of a to the base g, and written "$\operatorname{ind}_g a$". That is, if $(a, q) = 1$, then $\operatorname{ind}_g a$ will stand for any number t such that $g^t \equiv a \pmod q$. It is only determined $(\operatorname{mod} \varphi(q))$, since $a^{t+\varphi(q)} \equiv a^t \pmod q$. The following facts are immediate consequences of the definition.

THEOREM 4-10. If g is a primitive root of q then

$$\operatorname{ind}_g a \equiv \operatorname{ind}_g b \pmod{\varphi(q)} \quad \text{if} \quad a \equiv b \pmod q,$$

$$\operatorname{ind}_g (ab) \equiv \operatorname{ind}_g a + \operatorname{ind}_g b \pmod{\varphi(q)},$$

and

$$\operatorname{ind}_g a^n \equiv n \operatorname{ind}_g a \pmod{\varphi(q)}.$$

The procedure for finding the indices of the elements of a reduced residue system is quite simple if a primitive root is known. If g is a primitive root of q, construct a table of two rows and $\varphi(q)$ columns, of which the second row consists of the integers $1, 2, \ldots, \varphi(q)$, in order. In the first row enter g in the first column. Multiply this by g and reduce modulo q for the element in the second column; multiply this result by g and reduce modulo q for the element in the third column, etc. (When the table is complete, the last element in the first row should be 1.) Then the index of any element of the first row appears directly below that element.

If, for example, $q = 17$ and $g = 3$, we have the table

a:	3	9	10	13	5	15	11	16	14	8	7	4	12	2	6	1
ind a:	1	2	3	4	5	6	7	8	9	10	11	12	13	14	15	16

whereas for $q = 18$ and $g = 5$, we have

a:	5	7	17	13	11	1
ind a:	1	2	3	4	5	6

By Theorem 4–6, if $\text{ord}_m\, g = \varphi(m)$, then

$$\text{ord}_m\, g^n = \frac{\varphi(m)}{(n, \varphi(m))},$$

so that a is a primitive root of m if and only if $(\text{ind}\, a, \varphi(m)) = 1$. Thus in the above table we see that the primitive roots of 18 are 5 and 11, since the only numbers less than $\varphi(18) = 6$ and prime to it are 1 and 5.

Indices are quite useful in solving binomial congruences. For example, the congruence

$$10x \equiv 8 \pmod{18}$$

implies

$$5x \equiv 4 \pmod 9,$$

which in turn implies

$$\begin{aligned} \text{ind}\, 5 + \text{ind}\, x &\equiv \text{ind}\, 4 \pmod 6, \\ \text{ind}\, x &\equiv \text{ind}\, 4 - \text{ind}\, 5 \pmod 6. \end{aligned}$$

Since 2 is a primitive root of 9, we construct the table as before:

n:	2	4	8	7	5	1
ind n:	1	2	3	4	5	6

Thus

$$\text{ind}\, x \equiv 2 - 5 \equiv 3 \pmod 6,$$

whence

$$x \equiv 8 \pmod 9,$$

so that

$$x \equiv 8 \text{ or } 17 \pmod{18}.$$

We can also use indices to study the special polynomial congruence

$$x^n \equiv c \pmod p;$$

we have already considered the case $n = 2$ in the preceding chapter. This congruence is entirely equivalent to

$$n \cdot \text{ind}\, x \equiv \text{ind}\, c \pmod{p - 1},$$

which has solutions if and only if $(n, p - 1) | \text{ind}\, c$; if this condition is satisfied there are $d = (n, p - 1)$ roots. Such a criterion has the disadvantage that it requires knowledge of the value of ind c, and for this reason the following is more useful.

THEOREM 4–11. Let $(c, q) = 1$, where q is any number which has primitive roots. Then a necessary and sufficient condition for the congruence

$$x^n \equiv c \pmod{q} \tag{3}$$

to be solvable is that

$$c^{\varphi(q)/d} \equiv 1 \pmod{q},$$

where $d = (n, \varphi(q))$.

Proof: By an argument similar to that just given for prime modulus, a necessary and sufficient condition for the solvability of (3) is that ind $c \equiv 0 \pmod{d}$. This is equivalent to

$$\frac{\varphi(q)}{d} \text{ind}\, c \equiv 0 \pmod{\varphi(q)},$$

or, what is the same thing,

$$c^{\varphi(q)/d} \equiv 1 \pmod{q}. \ \blacktriangle$$

If $x^n \equiv c \pmod{m}$ is solvable and $(m, c) = 1$, then c is said to be an *nth-power residue* of m, otherwise a nonresidue.

THEOREM 4–12. The number of incongruent nth-power residues of q is $\varphi(q)/d$, and these residues are the roots of the congruence

$$x^{\varphi(q)/d} \equiv 1 \pmod{q}.$$

Proof: The second statement is a paraphrase of Theorem 4–11. Since q has a primitive root g, the roots of the congruence $x^{\varphi(q)/d} \equiv 1 \pmod{q}$ are the numbers g^t for which

$$g^{t\varphi(q)/d} \equiv 1 \pmod{q},$$

and this requires that $d|t$. But the number of multiples t of d with $1 \leq t \leq \varphi(q)$ is exactly $\varphi(q)/d$. (Note that this is a generalization of Theorem 3–15.) ▲

PROBLEMS

1. Given 2 as a primitive root of 29, construct a table of indices, and use it to solve the following congruences:

(a) $5x \equiv 21 \pmod{29}$,
(b) $17x \equiv 10 \pmod{29}$,
(c) $17x^2 \equiv 10 \pmod{29}$,
(d) $x^2 \equiv 20 \pmod{29}$,
(e) $x^2 - 4x - 16 \equiv 0 \pmod{29}$,
(f) $17x^2 - 3x + 10 \equiv 0 \pmod{29}$,
(g) $17x^2 - 4x + 1 \equiv 0 \pmod{29}$,
(h) $x^7 \equiv 17 \pmod{29}$.

2. Decide whether each of the following congruences is solvable:

(a) $x^5 \equiv 3 \pmod{31}$
(b) $x^3 - 3x^2 + 3x - 8 \equiv 0 \pmod{19}$.

3. Let q be a number having primitive roots. Show that h is a primitive root of q if and only if h is an rth power nonresidue of q for every prime r dividing $\varphi(q)$. [*Hint:* Write $h = g^k$, where g is a primitive root of q, and show that each of the allegedly equivalent statements is equivalent to the equation $(k, \varphi(q)) = 1$.] By eliminating all the appropriate powers of the elements of a reduced residue system, find all primitive roots of 13 and of 29. (Note the connection with Problem 3, Section 4–2.)

4. Show that if g and h are two different primitive roots of p, then

$$\operatorname{ind}_h a = \operatorname{ind}_g a \cdot \operatorname{ind}_h g \pmod{p-1}.$$

CHAPTER 5

CONTINUED FRACTIONS

5–1 Introduction. Much of the content of the preceding chapters depends, in the end, on the division theorem, Theorem 1–1. We now return to this theorem as the source of yet another important range of ideas in number theory. For convenience, we change the notation slightly. Let s and t be nonzero integers; then Theorem 1–1 asserts that there are unique integers a and r such that

$$s = ta + r, \qquad 0 \leq r < t. \tag{1}$$

It is useful now to describe the pair a, r by a condition on a rather than by the above inequality involving r, and to do so we write

$$\frac{s}{t} = a + \frac{r}{t}, \qquad 0 \leq \frac{r}{t} < 1. \tag{2}$$

We see from these relations that a must be chosen as the largest integer which does not exceed s/t, and conversely, if a is so chosen, then the difference $s/t - a$ is a fraction which is nonnegative and smaller than 1; its denominator is t and its numerator is the integer r of (1). The notion of the largest integer not exceeding a given real number x occurs repeatedly in the theory of numbers, and it has been dignified by a special notation: the largest integer not exceeding x is designated by $[x]$. In this notation, we see that the integers a and r satisfying (1) are $a = [s/t]$ and

$$r = t\left(\frac{s}{t} - \left[\frac{s}{t}\right]\right) = s - t\left[\frac{s}{t}\right].$$

The importance of this new way of looking at the division theorem lies in the possibility of generalizing (2) by allowing an arbitrary real number to replace the rational number s/t. That is, for every real number x we can write

$$x = a + x_1, \qquad 0 \leq x_1 < 1, \tag{3}$$

if we choose $a = [x]$, and (3) reduces to the division theorem when x is rational. In view of this, it is natural to ask whether there is also an analog of the Euclidean algorithm for real numbers. Returning to the rational case for a moment, we see that the Euclidean algorithm can be written in the following form:

$$\begin{aligned} \frac{s}{t} &= a_0 + \frac{r_0}{t}, & 0 < \frac{r_0}{t} < 1, \\ \frac{t}{r_0} &= a_1 + \frac{r_1}{r_0}, & 0 < \frac{r_1}{r_0} < 1, \\ \frac{r_0}{r_1} &= a_2 + \frac{r_2}{r_1}, & 0 < \frac{r_2}{r_1} < 1, \\ &\vdots \\ \frac{r_{N-2}}{r_{N-1}} &= a_N. \end{aligned} \tag{3}$$

In the first equation, s/t can be any rational number at all, so $a_0 = [s/t]$ can be any integer, positive, negative, or zero; but because of the inequalities, the remaining integers $a_1, a_2, \ldots$ are all positive. If we put $x = s/t$, $x_1 = t/r_0$, $x_2 = r_0/r_1, \ldots$, then we can write

$$\begin{aligned} x &= a_0 + \frac{1}{x_1}, & x_1 > 1, \\ x_1 &= a_1 + \frac{1}{x_2}, & x_2 > 1, \\ x_2 &= a_2 + \frac{1}{x_3}, & x_3 > 1, \\ &\vdots \end{aligned} \tag{4}$$

and this makes sense even if x is not a rational number. Of course, in this more general case there is no reason that the process must terminate, and in fact we see immediately that it cannot do so, since every x_n is then irrational and therefore never exactly equal to the integer a_n.

Let k be a positive integer which is otherwise unrestricted if x is irrational, but is smaller than N if $x = s/t$ is a rational number and the equations (3) hold. If from the first k of the equations (4) we eliminate $x_1, x_2, \ldots, x_{k-1}$, we obtain the relations

$$\begin{aligned} x &= a_0 + \frac{1}{x_1} \\ &= a_0 + \cfrac{1}{a_1 + \cfrac{1}{x_2}} = \cdots = a_0 + \cfrac{1}{a_1 + \cfrac{1}{a_2 + \ddots + \cfrac{1}{a_{k-1} + \cfrac{1}{x_k}}}}. \end{aligned} \tag{5}$$

On the other hand, if x is rational, equations (3) lead to the relation

$$x = a_0 + \cfrac{1}{a_1 + \cfrac{1}{a_2 + \cdots + \cfrac{1}{a_N}}}. \tag{6}$$

For example, starting with $x = \pi/4 = 0.785398\ldots$, we find

$$x = 0 + \frac{1}{1.273820\ldots},$$

$$1.273820\ldots = 1 + \frac{1}{3.65202\ldots},$$

$$3.65202\ldots = 3 + \frac{1}{1.53368\ldots},$$

$$1.53368\ldots = 1 + \frac{1}{1.8737\ldots},$$

$$\vdots$$

and hence

$$\frac{\pi}{4} = \cfrac{1}{1 + \cfrac{1}{3 + \cfrac{1}{1 + \cfrac{1}{1.8737\ldots}}}},$$

this being the expansion (5) with $k = 4$.

The complicated fractions occurring in equations (5) and (6) are called *continued fractions* or, more precisely, *finite* continued fractions, because there are only finitely many a_i. The latter integers are called *partial quotients*, while the numbers x_k are called *complete quotients*. A continued fraction such as (6) in which only integers appear, all of them except possibly a_0 being positive, is said to be *simple*. We shall be principally concerned with simple continued fractions, but to use (5) effectively we shall first prove some theorems concerning (6) which are valid whether or not the numbers a_i are integers, so long as all of them, except possibly a_0, are positive.

Problems

1. Find the simple continued fraction expansions of the following numbers: (a) 81/35, (b) 21/13, (c) 5, (d) −86/31, (e) 1/7.

2. Find the expansion (5), with $k = 4$, of the following numbers: (a) π, (b) $\sqrt{3}$, (c) $(1 + \sqrt{5})/2$, (d) 277/101.

3. Prove the following theorems concerning the greatest-integer function. Here and in the next problem, x and y are real numbers and n is an integer:

(a) $x = [x] + \theta$, where $0 \leq \theta < 1$;
(b) $x - 1 < [x] \leq x < [x] + 1$;
(c) $[x + n] = [x] + n$;
(d) $[[x]/n] = [x/n]$ for $n > 0$.

4. (a) Graph the following functions:
 (i) $y = [x]$,
 (ii) $y = [-x]$,
 (iii) $y = -[-x]$,
 (iv) $y = [2x]$.

(b) Show that $[x] + [y] \leq [x + y]$.
(c) Show that $-[-x]$ is the smallest integer not less than x.

5. The reduced fractions

$$\frac{p_0}{q_0} = \frac{a_0}{1}, \qquad \frac{p_1}{q_1} = a_0 + \frac{1}{a_1}, \qquad \frac{p_2}{q_2} = a_0 + \cfrac{1}{a_1 + \cfrac{1}{a_2}}, \ldots$$

are called the convergents of the continued fraction (6). Show, for the numerical examples of Problem 1, that always

$$\begin{vmatrix} p_n & q_n \\ p_{n+1} & q_{n+1} \end{vmatrix} = \pm 1.$$

6. Let x be a number between 0 and 1. Let a_1 be the smallest positive integer such that the difference

$$x_1 = x - \frac{1}{a_1}$$

is nonnegative, let a_2 be the smallest positive integer such that the difference

$$x_2 = x_1 - \frac{1}{a_2}$$

is nonnegative, etc. Show that this leads to a finite expansion

$$x = \frac{1}{a_1} + \frac{1}{a_2} + \cdots + \frac{1}{a_n}$$

(that is, that $x_{n+1} = 0$ for some n) if and only if x is rational.

7. (a) If m and n are positive integers, show that the number of multiples of m not exceeding n is $[n/m]$. (b) Let p be a positive prime and n a positive integer. Show that the power of p occurring in the prime decomposition of $n! = 1 \cdot 2 \cdot 3 \cdots n$ is

$$p^{[n/p]+[n/p^2]+[n/p^3]+\cdots}.$$

(c) Find the power of 2 occurring in 10!, and also the power of 5. With how many zeros does the decimal expansion of 10! end? of 100!?

5-2 The basic identities. Let $z_0, z_1, \ldots, z_k$ be real numbers, all of which, except possibly the first, are positive, and consider the continued fraction

$$x = z_0 + \cfrac{1}{z_1 + \cfrac{1}{z_2 + \ddots + \cfrac{1}{z_{k-1} + \cfrac{1}{z_k}}}}. \tag{7}$$

Now clearly x is determined completely by the z's, so we shall abbreviate the cumbersome equation (7) by writing $x = \{z_0; z_1, \ldots, z_k\}$. The reason for the semicolon in this notation is to emphasize the distinction between (7) and the continued fraction

$$\cfrac{1}{z_0 + \cfrac{1}{z_1 + \cfrac{1}{z_2 + \ddots + \cfrac{1}{z_k}}}} = \{0; z_0, z_1, \ldots, z_k\};$$

moreover, the number preceding the semicolon plays a rather different role from the other z's in that it can be zero or negative. By placing parentheses around the fraction $z_{k-1} + 1/z_k$ at the bottom of (7), we see that

$$\{z_0; z_1, \ldots, z_{k-1}, z_k\} = \left\{z_0; z_1, \ldots, z_{k-2}, z_{k-1} + \frac{1}{z_k}\right\}. \tag{8}$$

The continued fractions

$$\{z_0;\}, \{z_0; z_1\}, \{z_0; z_1, z_2\}, \ldots, \{z_0; z_1, z_2, \ldots, z_k\}$$

are called the *convergents* of the expansion (7). If we simplify the first few to ordinary fractions, we obtain

$$\begin{aligned}
\{z_0;\} &= \frac{z_0}{1},\\
\{z_0; z_1\} &= \frac{z_0 z_1 + 1}{z_1},\\
\{z_0; z_1, z_2\} &= \frac{z_0 z_1 z_2 + z_0 + z_2}{z_1 z_2 + 1},\\
&\vdots
\end{aligned}$$

We define the numbers p_n and q_n, for $n = 1, \ldots, k$, as being the numerators and denominators of the fractions just written, so that

$$\begin{aligned} &p_0 = z_0, \quad q_0 = 1, \\ &p_1 = z_0 z_1 + 1, \quad q_1 = z_1, \\ &p_2 = z_0 z_1 z_2 + z_0 + z_2, \quad q_2 = z_1 z_2 + 1, \\ &\vdots \end{aligned} \tag{9}$$

and refer to p_n and q_n as the numerator and denominator of the nth convergent of (7). (Note that this is a genuine definition, because the ratio of $2z_0$ to 2 is the number $\{z_0;\}$; but, according to this definition, $2z_0$ and 2 are not the numerator and denominator of $\{z_0;\}$.

Returning to the numerical example given in the preceding section, we have

$$z_0 = 0, \quad z_1 = 1, \quad z_2 = 3, \quad z_3 = 1, \quad z_4 = 1.8737\ldots,$$

and hence

$$\begin{aligned} &p_0 = 0, && q_0 = 1, && \frac{p_0}{q_0} = 0, \\ &p_1 = 1, && q_1 = 1, && \frac{p_1}{q_1} = 1, \\ &p_2 = 3, && q_2 = 4, && \frac{p_2}{q_2} = \frac{3}{4}, \\ &p_3 = 4, && q_3 = 5, && \frac{p_3}{q_3} = \frac{4}{5}, \\ &p_4 = 4z_4 + 3, && q_4 = 5z_4 + 4, && \frac{p_4}{q_4} = \frac{4z_4 + 3}{5z_4 + 4} = \frac{\pi}{4}. \end{aligned}$$

THEOREM 5–1. *The numerators p_n and the denominators q_n of the nth convergent of (7) satisfy the equations*

$$p_0 = z_0, \quad p_1 = z_0 z_1 + 1, \quad p_n = p_{n-1} z_n + p_{n-2} \quad \text{for} \quad 2 \le n \le k,$$

and

$$q_0 = 1, \quad q_1 = z_1, \quad q_n = q_{n-1} z_n + q_{n-2} \quad \text{for} \quad 2 \le n \le k. \tag{10}$$

In particular,

$$x = \frac{p_{k-1} z_k + p_{k-2}}{q_{k-1} z_k + q_{z-2}} \quad \text{if} \quad k \ge 2. \tag{11}$$

Proof: Equation (11) follows from (10) by taking $n = k$ and noting that $x = p_k/q_k$, so we need only prove (10). This we do by induction on n. According to (9) we have, for $n = 2$,

$$p_1 z_2 + p_0 = z_2(z_0 z_1 + 1) + z_0 = p_2$$

and, similarly, $q_1z_2 + q_0 = q_2$. Now suppose that the equations involving n in (10) are correct, and that $n < k$. Using the principle illustrated in (8), we obtain

$$\frac{p_{n+1}}{q_{n+1}} = \{z_0; z_1, \ldots, z_{n+1}\} = \left\{z_0; z_1, \ldots, z_n + \frac{1}{z_{n+1}}\right\}$$

$$= \frac{p_{n-1}\left(z_n + \dfrac{1}{z_{n+1}}\right) + p_{n-2}}{q_{n-1}\left(z_n + \dfrac{1}{z_{n+1}}\right) + q_{n-2}} = \frac{(p_{n-1}z_n + p_{n-2}) + \dfrac{p_{n-1}}{z_{n+1}}}{(q_{n-1}z_n + q_{n-2}) + \dfrac{q_{n-1}}{z_{n+1}}}$$

$$= \frac{p_n + \dfrac{p_{n-1}}{z_{n+1}}}{q_n + \dfrac{q_{n-1}}{z_{n+1}}} = \frac{p_nz_{n+1} + p_{n-1}}{q_nz_{n+1} + q_{n-1}},$$

and this gives the required expressions for p_{n+1} and q_{n+1}. ▲

One says that the sequences $p_0, p_1, p_2, \ldots$ and $q_0, q_1, q_2, \ldots$ are defined *recursively* by equations (10), because each element after the second in each sequence is defined in terms of earlier elements. The Fibonacci sequence discussed in Section 1-3 was also defined recursively.

THEOREM 5-2. We have, for $1 \leq n \leq k$,

$$p_nq_{n-1} - p_{n-1}q_n = (-1)^{n-1}, \tag{12}$$

or equivalently,

$$\frac{p_n}{q_n} - \frac{p_{n-1}}{q_{n-1}} = \frac{(-1)^{n-1}}{q_nq_{n-1}}.$$

Proof: We again proceed by induction on n. Equation (12) holds for $n = 1$, by (9). For $n > 1$ we see from (10) that if

$$p_{n-1}q_{n-2} - p_{n-2}q_{n-1} = (-1)^{n-2},$$

then

$$\begin{aligned} p_nq_{n-1} - p_{n-1}q_n &= (p_{n-1}z_n + p_{n-2})q_{n-1} - p_{n-1}(q_{n-1}z_n + q_{n-2}) \\ &= p_{n-2}q_{n-1} - p_{n-1}q_{n-2} = -(-1)^{n-2} \\ &= (-1)^{n-1}. \ ▲ \end{aligned}$$

THEOREM 5-3. For $2 \leq n \leq k$,

$$\frac{p_n}{q_n} - \frac{p_{n-2}}{q_{n-2}} = \frac{(-1)^n z_n}{q_nq_{n-2}}.$$

The proof of this theorem follows exactly the same lines as that of Theorem 5–2, and is left to the reader.

THEOREM 5–4. The convergents are related to each other and to x by the following inequalities:

$$\frac{p_0}{q_0} < \frac{p_2}{q_2} < \cdots < x < \cdots < \frac{p_3}{q_3} < \frac{p_1}{q_1},$$

except that the last convergent, p_N/q_N, is equal to x. That is, the even convergents, p_0/q_0, $p_2/q_2, \ldots$, form a strictly increasing sequence of numbers, with none larger than x, and the odd convergents, p_1/q_1, $p_3/q_3, \ldots$, form a strictly decreasing sequence, with none smaller than x.

Proof: From (10) we see that all q's are positive. Hence it follows from Theorems 5–2 and 5–3 that the differences

$$\frac{p_n}{q_n} - \frac{p_{n-1}}{q_{n-1}} \quad \text{and} \quad \frac{p_n}{q_n} - \frac{p_{n-2}}{q_{n-2}}$$

have opposite signs, which means that each convergent lies between the two preceding ones. It is clear that $p_0/q_0 < p_1/q_1$, so we successively obtain

$$\frac{p_0}{q_0} < \frac{p_2}{q_2} < \frac{p_1}{q_1},$$

$$\frac{p_0}{q_0} < \frac{p_2}{q_2} < \frac{p_3}{q_3} < \frac{p_1}{q_1},$$

$$\frac{p_0}{q_0} < \frac{p_2}{q_2} < \frac{p_4}{q_4} < \frac{p_3}{q_3} < \frac{p_1}{q_1},$$

and so on. Since x itself is either the largest of the even convergents or the smallest of the odd convergents, the theorem follows. ▲

The recursion relations (10) provide a simple procedure for actually calculating the successive convergents of a continued fraction. Consider for example the continued fraction $\{3; 1, 4, 2, 7\}$. We construct the following table:

n:	0	1	2	3	4
a_n:	3	1	4	2	7
p_n:	3	4			
q_n:	1	1			

Here p_0, p_1, q_0, q_1 have been computed in accordance with (10). To determine p_2 and q_2, we multiply a_2 by p_1 and add p_0, obtaining 19, and we multiply a_2 by q_1 and add q_0, obtaining 5, as indicated by the dotted lines. Continuing in this fashion, we complete the table:

n:	0	1	2	3	4
a_n:	3	1	4	2	7
p_n:	3	4	19	42	313
q_n:	1	1	5	11	82

Thus $\{3; 1, 4, 2, 7\} = 313/82$, and the convergents are 3/1, 4/1, 19/5, 42/11, 313/82.

It should perhaps be mentioned that if we define

$$p_{-2} = 0, \qquad p_{-1} = 1,$$
$$q_{-2} = 1, \qquad q_{-1} = 0,$$

then equations (10) can be written more simply as

$$\left.\begin{aligned} p_n &= p_{n-1}z_n + p_{n-2} \\ q_n &= q_{n-1}z_n + q_{n-2} \end{aligned}\right\} \quad \text{for} \quad n \geq 0.$$

This also simplifies the construction of tables of convergents such as that above, since it is no longer necessary to work out p_0/q_0 and p_1/q_1 separately. For example:

n:	−2	−1	0	1	2
a_n:			3	1	4
p_n:	0	1	3	4	19
q_n:	1	0	1	1	5

Problems

1. Compute the convergents of the following continued fractions:
(a) $\{3; 7, 2, 1, 1, 2\}$,
(b) $\{1; 2, 3, 4, 5\}$,
(c) $\{1; 1, 1, 1, 1\}$. (What if the 1's were continued further?)

2. Find the continued fraction expansions of 3.14159 and 3.1416. What can you say about a continued fraction expansion for π?

3. Suppose that all numbers z_n in (10) are positive integers. Then show that for $n \geq 0$, we have $(p_n, q_n) = 1$.

4. Prove Theorem 5–3.

5. Show that $q_n/q_{n-1} = \{z_n; z_{n-1}, \ldots, z_1\}$.

6. Show that $p_n/p_{n-1} = \{z_n; z_{n-1}, \ldots, z_1, z_0\}$ if $z_0 > 0$, whereas $p_n/p_{n-1} = \{z_n; z_{n-1}, \ldots, z_2\}$ if $z_0 = 0$.

7. If there is any sense to be attached to the equation $x = \{1; 1, 1, 1, \dots\}$, then clearly $x = \{1; x\}$. Use this to find the only possible value for x. Similarly, find the only possible value of $\{2; 3, 2, 3, 2, 3, \dots\}$, and of $\{3; 4, 1, 4, 1, 4, 1, \dots\}$. Can you make (and prove) a general statement about the values of all nonterminating simple continued fractions in which the z_k's form periodic sequences of positive integers?

5–3 The simple continued fraction expansion of a rational number. We now return to the simple continued fractions, in which the partial quotients are positive integers, and consider first the expansion of a rational number as such a fraction. We have seen that by eliminating the remainders in the Euclidean algorithm, we obtain the finite expansion (6), so that every rational number has a finite simple continued fraction expansion. There remains the possibility that there are several such expansions for a single number, and it is even conceivable that there is a nonterminating simple continued fraction which in some sense represents the number, just as the nonterminating decimal 0.333... represents 1/3. The latter possibility will be eliminated in the next section; for the moment we restrict attention to the finite case.

THEOREM 5–5. *There is only one finite simple continued fraction $\{a_0; a_1, \dots, a_N\}$ whose value is a specified rational number x, if it is required that a_N be larger than 1. The only other finite simple continued fraction with value x is $\{a_0; a_1, \dots, a_N - 1, 1\}$.*

The ambiguity described is illustrated by the following expansions of 4/11:

$$\frac{4}{11} = \cfrac{1}{2 + \cfrac{1}{1 + \cfrac{1}{3}}} = \cfrac{1}{2 + \cfrac{1}{1 + \cfrac{1}{2 + \cfrac{1}{1}}}}.$$

The idea of the proof is very simple. Since $\{0; a_1, \dots, a_N\}$ is a number between 0 and 1 for all $a_1, \dots, a_N$, the integer a_0 such that

$$x = \{a_0; a_1, \dots, a_N\}$$

for suitable $a_1, \dots, a_N$ is uniquely determined: it is $[x]$. Thus any two expansions of x agree in the first partial quotient, and we can subtract this common part and go on to the next partial quotient, where the uniqueness argument can be repeated. A formal inductive proof follows.

Proof: Suppose that $x = \{a_0; a_1, \dots, a_N\} = \{b_0; b_1, \dots, b_M\}$, and that $a_N > 1$, $b_M > 1$. If we set $a'_n = \{a_n; a_{n+1}, \dots, a_N\}$, then in analogy with (8) we can write

$$x = \{a_0; a_1, \ldots, a_{n-1}, a_n'\}, \qquad 1 \le n \le N, \tag{13}$$

and similarly, for $b_n' = \{b_n; b_{n+1}, \ldots, b_M\}$, we have

$$x = \{b_0; b_1, \ldots, b_{n-1}, b_n'\}, \qquad 1 \le n \le M. \tag{14}$$

For $n < N$, we obtain

$$a_n' = a_n + \frac{1}{a_{n+1}'}, \qquad a_{n+1}' > 1, \tag{15}$$

so that $a_n = [a_n']$. Similarly, for $n < M$, we find $b_n = [b_n']$. Taking $n = 0$, we have $a_0' = b_0' = x$, so that $a_0 = b_0 = [x]$. We proceed by induction. Suppose that

$$a_0 = b_0, \quad a_1 = b_1, \ldots, \quad a_{n-1} = b_{n-1}.$$

Then if $p_{n-1}/q_{n-1} = \{a_0; a_1, \ldots, a_{n-1}\}$, we find by (11), (13), and (14) that

$$x = \frac{p_{n-1}a_n' + p_{n-2}}{q_{n-1}a_n' + q_{n-2}} = \frac{p_{n-1}b_n' + p_{n-2}}{q_{n-1}b_n' + q_{n-2}},$$

whence

$$(p_{n-1}a_n' + p_{n-2})(q_{n-1}b_n' + q_{n-2}) = (p_{n-1}b_n' + p_{n-2})(q_{n-1}a_n' + q_{n-2}),$$
$$(p_{n-2}q_{n-1} - p_{n-1}q_{n-2})b_n' = (p_{n-2}q_{n-1} - p_{n-1}q_{n-2})a_n';$$

thus, by (12), $b_n' = a_n'$. But then $a_n = [a_n'] = [b_n'] = b_n$. Hence all partial quotients are the same, and when one expansion terminates, so does the other.

This argument breaks down if any $a_{n+1}' = 1$, since that possibility was excluded from (15). If in fact $a_{n+1}' = 1$, then $a_n = [a_n'] - 1$, $a_{n+1} = 1$, and the expansion terminates at this point. Therefore the preceding complete quotients a_k' were larger than 1 (otherwise the expansion would have terminated earlier), and the preceding partial quotients a_k and b_k were equal. ▲

According to Theorem 5-5, a rational number s/t has two finite simple continued fraction expansions $\{a_0; a_1, \ldots, a_N\}$, and in one of them N is even and in the other N is odd. If we apply Theorem 5-2 with $n = N$, we have $p_N/q_N = s/t$ and $q_{N-1}s - p_{N-1}t = (-1)^{N-1}$. Thus we arrive at

THEOREM 5-6. The linear diophantine equation $sx - ty = 1$ has the solution $x = q_{N-1}$ and $y = p_{N-1}$ if $s/t = \{a_0; a_1, \ldots, a_N\}$ and N is odd.

When s and t are large, this is probably the quickest method of solving the linear diophantine equation.

Problem

1. Use the method of this section to solve the following diophantine equations:
(a) $3154x - 2971y = 1$,
(b) $3154x + 2971y = 45$,
(c) $31416x + 10000y = 8$.

5–4 The expansion of an irrational number. We have seen that for irrational x the algorithm (4) leads to the equations (5), and the latter are of precisely the same form as (13), if $a_n' = x_n$. For irrational x, the complete quotients x_n are always larger than 1, so that the argument following (13) shows that if for arbitrary n,

$$x = \{a_0; a_1, \ldots, a_n, x_{n+1}\} = \{b_0; b_1, \ldots, b_n, x_{n+1}'\},$$

where the a_k and b_k are integers, and all—with the possible exception of a_0 and b_0—are positive, then $a_k = b_k$ for $k = 1, \ldots, n$. If the algorithm (4) is continued indefinitely, an *infinite* simple continued fraction $\{a_0; a_1, a_2, \ldots\}$ results, and there is just one such fraction, corresponding to each irrational number x. The question we must now face is, what sense can be made of the equation $x = \{a_0; a_1, a_2, \ldots\}$?

Theorem 5–7. If the infinite simple continued fraction $\{a_0; a_1, a_2, \ldots\}$ is associated with x by means of equations (4), and this continued fraction has convergents $p_0/q_0, p_1/q_1, \ldots$, then $\lim_{n\to\infty} p_n/q_n = x$. That is, the difference between x and the nth convergent approaches 0 as n increases without limit.

Proof: If we put $x = \{a_0; a_1, \ldots, a_n, x_{n+1}\}$, then we obtain from (11)

$$x - \frac{p_n}{q_n} = \frac{p_n x_{n+1} + p_{n-1}}{q_n x_{n+1} + q_{n-1}} - \frac{p_n}{q_n} = \frac{(-1)^n}{q_n(q_n x_{n+1} + q_{n-1})}, \tag{16}$$

and hence, since q_n increases without bound as $n \to \infty$,

$$\lim_{n\to\infty} \left(x - \frac{p_n}{q_n}\right) = 0. \blacktriangle$$

If the sequence of convergents of an infinite continued fraction converges to a certain number x, we say that the value of the continued fraction is x, or that the continued fraction converges to, or equals, x. What we have just shown is that to every irrational number x, there corresponds a unique infinite simple continued fraction whose value is x, and that this continued fraction is generated by the algorithm (4). The following theorem gives the complementary result.

THEOREM 5-8. Every infinite simple continued fraction converges to a real number.

Proof: Let the continued fraction be $\{a_0; a_1, a_2, \ldots\}$, with convergents p_1/q_1, $p_2/q_2, \ldots$, and let X be the rational number p_n/q_n. Then X has the expansion $\{a_0; a_1, \ldots, a_n\}$, and this finite continued fraction has the same convergents p_k/q_k as the infinite expansion, for $k = 1, \ldots, n$. Thus the inequalities of Theorem 5-4 hold in this same range (with x replaced by X), and we deduce that the even convergents of the infinite expansion form an increasing sequence bounded above by p_1/q_1, for example, and the odd convergents form a decreasing sequence bounded below by p_0/q_0. A fundamental principle concerning infinite sequences of real numbers is that every increasing sequence which is bounded above is convergent, and that every decreasing sequence which is bounded below is convergent. Hence the limits

$$\lim_{n\to\infty} \frac{p_{2n}}{q_{2n}} \quad \text{and} \quad \lim_{n\to\infty} \frac{p_{2n+1}}{q_{2n+1}}$$

exist, and to prove that $\lim_{n\to\infty} p_n/q_n$ exists, it suffices to show that they are equal. To this end we invoke Theorem 5-2, with $2n$ in place of n:

$$\frac{p_{2n}}{q_{2n}} - \frac{p_{2n-1}}{q_{2n-1}} = \frac{-1}{q_{2n}q_{2n-1}}.$$

From equations (10) we see that

$$q_0 = 1, \qquad q_1 \geq 1, \qquad q_n \geq q_{n-1} + q_{n-2},$$

and from these relations it is easy to prove by induction that $q_n \geq n$ for $n = 1, 2, \ldots$. It follows from (15) that

$$\lim_{n\to\infty} \left(\frac{p_{2n}}{q_{2n}} - \frac{p_{2n-1}}{q_{2n-1}}\right) = 0,$$

and since the separate limits are known to exist, we have

$$\lim_{n\to\infty} \frac{p_{2n}}{q_{2n}} = \lim_{n\to\infty} \frac{p_{2n-1}}{q_{2n-1}}.$$

Hence $\lim p_n/q_n$ exists. ▲

THEOREM 5-9. The simple continued fraction expansion of a rational number is always finite. Equivalently, the value of an infinite simple continued fraction is always irrational.

Proof: We have seen that an infinite simple continued fraction converges to a real number x, and that the continued fraction results by applying the algorithm (4) to this number x. If x were rational, the expansion would be finite, since it is then just the Euclidean algorithm. ▲

THEOREM 5–10. If x is irrational, the sequence $\{p_{2n}/q_{2n}\}$ is an increasing sequence with limit x, and the sequence $\{p_{2n+1}/q_{2n+1}\}$ is a decreasing sequence with limit x. Moreover, each convergent is closer to x than the preceding one:

$$\left|x - \frac{p_n}{q_n}\right| < \left|x - \frac{p_{n-1}}{q_{n-1}}\right|, \quad \text{for} \quad n \geq 1.$$

Proof: The first sentence of the theorem is simply a combination of Theorems 5–4 and 5–8. Using (16), we have

$$|q_n x - p_n| = \frac{1}{q_n x_{n+1} + q_{n-1}}.$$

Since x is irrational, $x_{n+1} > [x_{n+1}] = a_{n+1}$, and hence

$$q_n x_{n+1} + q_{n-1} > q_n a_{n+1} + q_{n-1} = q_{n+1},$$

whereas

$$q_n x_{n+1} + q_{n-1} < q_n(a_{n+1} + 1) + q_{n-1} = q_{n+1} + q_n \leq q_{n+2}.$$

Thus

$$\frac{1}{q_{n+2}} < |q_n x - p_n| < \frac{1}{q_{n+1}}.$$

Since the q_n form an increasing sequence, it follows that

$$|q_n x - p_n| < |q_{n-1} x - p_{n-1}|, \tag{17}$$

and this is a stronger inequality than that of the theorem. ▲

PROBLEMS

1. Reconsider Problem 7 of Section 5–2.

2. Show that the continued fraction expansion of $\sqrt{3}$ is periodic, and compute the first few convergents. Proceed similarly for $\sqrt{13}$.

3. Is there any sense to be made of the equation

$$\frac{1 + \sqrt{5}}{2} = \sqrt{1 + \sqrt{1 + \sqrt{1 + \cdots}}}\ ?$$

4. Prove that (17) implies the inequality of Theorem 5–10.

5. What information can you get about the Fibonacci numbers from the equation $(1 + \sqrt{5})/2 = \{1; 1, 1, 1, \ldots\}$? How is it connected with the inequality $u_n < (7/4)^n$ proved in Chapter 1? [*Hint:* Compute a number of convergents.]

6. Describe by inequalities the set of real numbers having a fixed set of integers $a_0, a_1, \ldots, a_n$ as their first $n + 1$ partial quotients.

7. For $x > 1$, show that the kth convergent of the expansion of $1/x$ is the reciprocal of the $(k - 1)$-convergent of the expansion of x.

5-5 The expansion of quadratic irrationalities.

Decimal expansions of rational numbers are always either terminating or periodic, and we have just seen that simple continued fraction expansions of rational numbers always terminate. We shall now see that the infinite periodic continued fractions correspond exactly to the real quadratic irrational numbers, these being the real irrational numbers which are solutions of quadratic equations $ax^2 + bc + c = 0$ with integral coefficients a, b, c. According to the quadratic formula, such numbers are of the form $x + y\sqrt{d}$, where x and y are rational and d is a positive integer, not a square.

Consider for example the number $\xi = \sqrt{7}$. Designating the complete quotients by $\xi_1, \xi_2, \ldots$, we have

$$\sqrt{7} = 2 + (\sqrt{7} - 2), \qquad a_0 = 2, \quad \xi_1 = (\sqrt{7} - 2)^{-1};$$

$$\frac{1}{\sqrt{7} - 2} = \frac{\sqrt{7} + 2}{3} = 1 + \frac{\sqrt{7} - 1}{3}, \qquad a_1 = 1, \quad \xi_2 = \left(\frac{\sqrt{7} - 1}{3}\right)^{-1},$$

$$\frac{3}{\sqrt{7} - 1} = \frac{\sqrt{7} + 1}{2} = 1 + \frac{\sqrt{7} - 1}{2}, \qquad a_2 = 1, \quad \xi_3 = \left(\frac{\sqrt{7} - 1}{2}\right)^{-1},$$

$$\frac{2}{\sqrt{7} - 1} = \frac{\sqrt{7} + 1}{3} = 1 + \frac{\sqrt{7} - 2}{3}, \qquad a_3 = 1, \quad \xi_4 = \left(\frac{\sqrt{7} - 2}{3}\right)^{-1},$$

$$\frac{3}{\sqrt{7} - 2} = \sqrt{7} + 2 = 4 + (\sqrt{7} - 2), \qquad a_4 = 4, \quad \xi_5 = (\sqrt{7} - 2)^{-1}.$$

Since $\xi_5 = \xi_1$, also $\xi_6 = \xi_2$, $\xi_7 = \xi_3, \ldots$, so the sequence $\{\xi_k\}$ (and therefore also $\{a_k\}$) is periodic. Thus we have the periodic expansion

$$\sqrt{7} = \{2; 1, 1, 1, 4, 1, 1, 1, 4, \ldots\}.$$

Using the relations (10), we construct the following table:

k	0	1	2	3	4	5	6	$\cdots$
a_k	2	1	1	1	4	1	1	$\cdots$
p_k	2	3	5	8	37	45	82	$\cdots$
q_k	1	1	2	3	14	17	31	$\cdots$

Here the element $37 = p_4$, for example, is determined by multiplying $a_4 = 4$ by $p_3 = 8$ and adding $p_2 = 5$. Thus the convergents to $\sqrt{7}$ are $3, 5/2, 8/3, 37/14, 45/17, \ldots$.

Conversely, consider the continued fraction

$$\xi = \{1; 3, 1, 2, 1, 2, \ldots\},$$

where $a_{2n} = 1$ and $a_{2n+1} = 2$ for $n \geq 1$. We have

$$\xi_2 = \{1; 2, 1, 2, \ldots\} = \{1; 2, \xi_2\},$$

so that

$$\xi_2 = 1 + \cfrac{1}{2 + \cfrac{1}{\xi_2}} = 1 + \frac{\xi_2}{2\xi_2 + 1} = \frac{3\xi_2 + 1}{2\xi_2 + 1},$$

$$2\xi_2^2 - 2\xi_2 - 1 = 0,$$

$$\xi_2 = \frac{-1 + \sqrt{3}}{2}.$$

(The plus sign is chosen before the radical since $\xi_2 > 0$.) Hence

$$\xi = 1 + \cfrac{1}{3 + \cfrac{1}{\cfrac{\sqrt{3} - 1}{2}}} = \frac{4\sqrt{3} - 2}{3\sqrt{3} - 1} = \frac{17 - \sqrt{3}}{13}.$$

We can now show that these are not isolated phenomena.

Theorem 5–11. Every eventually periodic simple continued fraction converges to a quadratic irrationality, and every quadratic irrationality has a simple continued fraction expansion which is eventually periodic.

Proof: The first part is quite simple. Suppose that the first period begins with a_n, and let the length of the period be h; then $a_{k+h} = a_k$ for $k \geq n$. Set

$$\xi = \{a_0; a_1, \ldots\} \quad \text{and} \quad \xi_k = \{a_k; a_{k+1}, \ldots\},$$

so that $\xi_{k+h} = \xi_k$ for $k \geq n$. By this and equation (11),

$$\xi = \frac{p_{n-1}\xi_n + p_{n-2}}{q_{n-1}\xi_n + q_{n-2}} = \frac{p_{n+h-1}\xi_n + p_{n+h-2}}{q_{n+h-1}\xi_n + q_{n+h-2}},$$

and hence ξ_n satisfies a quadratic equation with integral coefficients. Since ξ_n is obviously not rational, it is a quadratic irrationality. Again by

(11), the same is true of ξ itself, since if

$$A\xi_n^2 + B\xi_n + C = 0,$$

then

$$A(-q_{n-2}\xi + p_{n-2})^2 + B(-q_{n-2}\xi + p_{n-2})(q_{n-1}\xi - p_{n-1}) + C(q_{n-1}\xi - p_{n-1})^2 = 0,$$

and this is a quadratic equation in ξ.

The proof of the converse involves a little more computation. Suppose that

$$f(\xi) = A\xi^2 + B\xi + C = 0,$$

where A, B, and C are integers, and ξ is irrational. Then equation (11) yields

$$A(p_{k-1}\xi_k + p_{k-2})^2 + B(p_{k-1}\xi_k + p_{k-2})(q_{k-1}\xi_k + q_{k-2}) + C(q_{k-1}\xi_k + q_{k-2})^2 = 0,$$

or

$$A_k\xi_k^2 + B_k\xi_k + C_k = 0,$$

where the integers A_k, B_k, and C_k are given by the equations

$$A_k = Ap_{k-1}^2 + Bp_{k-1}q_{k-1} + Cq_{k-1}^2,$$

$$B_k = 2Ap_{k-1}p_{k-2} + B(p_{k-1}q_{k-2} + p_{k-2}q_{k-1}) + 2Cq_{k-1}q_{k-2},$$

$$C_k = Ap_{k-2}^2 + Bp_{k-2}q_{k-2} + Cq_{k-2}^3.$$

Thus

$$A_k = q_{k-1}^2 f\left(\frac{p_{k-1}}{q_{k-1}}\right) \quad \text{and} \quad C_k = q_{k-2}^2 f\left(\frac{p_{k-2}}{q_{k-2}}\right).$$

We now use the identity

$$au^2 + bu + c = a(u - v)^2 + (2av + b)(u - v) + (av^2 + bv + c),$$

which is easily verified by multiplying out on the right and collecting terms.* Choosing $u = p_{k-1}/q_{k-1}$ and $v = \xi$, and using the fact that $a\xi^2 + b\xi + c = 0$, we obtain

$$A_k = q_{k-1}^2 \left(\frac{p_{k-1}}{q_{k-1}} - \xi\right) \left\{2a\xi + b + a\left(\frac{p_{k-1}}{q_{k-1}} - \xi\right)\right\}.$$

---

* This is also, of course, just the Taylor expansion of $f(u)$ near the point v.

Now by (16),

$$\left| \xi - \frac{p_{k-1}}{q_{k-1}} \right| = \frac{1}{q_{k-1}(q_{k-1}\xi_k + q_{k-2})} < \frac{1}{q_{k-1}^2}, \tag{18}$$

so that

$$|A_k| < |2a\xi + b| + \frac{|a|}{q_{k-1}^2},$$

and similarly,

$$|C_k| < |2a\xi + b| + \frac{|a|}{q_{k-2}^2}.$$

Thus $|A_k|$ and $|C_k|$ remain bounded as $k \to \infty$.

To see that $|B_k|$ is also bounded, we use the fact that all the quantities $B_k^2 - 4A_kC_k$ have the common value $B^2 - 4AC = D$. (This can be proved by a straightforward but tedious computation or, if one is acquainted with the theory of linear transformations, by noting that the expression $A_kx'^2 + B_kx'y' + C_ky'^2$ is obtained from $Ax^2 + Bxy + Cy^2$ by the unimodular substitution

$$x = p_{k-1}x' + p_{k-2}y', \qquad y = q_{k-1}x' + q_{k-2}y',$$

and that two such forms have the same discriminant.) Since A_k and C_k are bounded and D is fixed,

$$B_k^2 = D + 4A_kC_k$$

must be bounded also.

Thus, there is a constant M such that

$$|A_k| < M, \qquad |B_k| < M, \qquad |C_k| < M$$

for all k. Since there are fewer than $(2M + 1)^3$ triples of integers each numerically smaller than M, there must be three indices, say n_1, n_2, and n_3, which yield the same triple:

$$A_{n_1} = A_{n_2} = A_{n_3}, \qquad B_{n_1} = B_{n_2} = B_{n_3}, \qquad C_{n_1} = C_{n_2} = C_{n_3}.$$

Since the equation $A_{n_1}x^2 + B_{n_1}x + C_{n_1} = 0$ has only two roots, two of the numbers ξ_{n_1}, ξ_{n_2}, ξ_{n_3} must be equal; with suitable notation, they can be taken to be ξ_{n_1} and ξ_{n_2}, where $n_1 < n_2$. If $n_2 - n_1 = h$, then $\xi_{n_1+h} = \xi_{n_1}$, and

$$\xi_{n_1+h+1} = \frac{1}{\xi_{n_1+h} - [\xi_{n_1+h}]} = \frac{1}{\xi_{n_1} - [\xi_{n_1}]} = \xi_{n_1+1},$$

$$\xi_{n_1+h+2} = \frac{1}{\xi_{n_1+h+1} - [\xi_{n_1+h+1}]} = \frac{1}{\xi_{n_1+1} - [\xi_{n_1+1}]} = \xi_{n_1+2},$$

and, in general, $\xi_{k+h} = \xi_k$ for $k \geq n_1$. Thus the ξ_k's are eventually periodic. Since each a_k is determined exclusively by the corresponding ξ_k, the same is true of the a_k's, and the proof is complete. ▲

Problems

1. Find the continued fraction expansions of $\sqrt{d}$ for $d = 2, 3, 5, 6, 7, 8, 10$. Make all possible general conjectures which are consistent with these data and which seem reasonable to you, and test them against the cases $d = 11, 12, 13, 14$. Can you prove any of your conjectures?

2. The equation $x^2 - x - 1 = 0$ can be solved by transposing and dividing by x:

$$x = \frac{x+1}{x} = 1 + \frac{1}{x} = 1 + \frac{1}{1 + (1/x)} = \{1; 1, 1, 1, \ldots\} = \frac{1 + \sqrt{5}}{2}.$$

Solve the equation $ax^2 - abx - 1 = 0$, in which a and b are positive integers, by continued fractions. Try to find other equations which lend themselves to this approach.

3. Find the continued fraction expansions of $\sqrt{n^2 + 1}$ and $\sqrt{n^2 + n}$, where n is a positive integer.

4. Prove the assertion made in the text that for $k \geq 2$,

$$B_k^2 - 4A_kC_k = B^2 - 4AC.$$

5. Below is an outline of a proof that the expansion of $\sqrt{d}$ (d a positive non-square integer) is periodic after a_0. Fill in all details. (If $\alpha = r + s\sqrt{d}$, where r and s are rational, then $\bar{\alpha} = r - s\sqrt{d}$.)

Put $\xi = \sqrt{d} + [\sqrt{d}]$. Then $-1 < \bar{\xi} < 0$, and from the equation

$$\xi_k = a_k + \frac{1}{\xi_{k+1}}$$

it follows that $-1 < \bar{\xi}_k < 0$ for $k \geq 1$. This in turn shows that $a_k = [-1/\bar{\xi}_{k+1}]$. Now suppose that the periodicity of $\{\xi_k\}$ begins when $k = n$, and that the period is of length h, so that $\xi_n = \xi_{n+h}$. Consequently, $a_{n-1} = a_{n+h-1}$, and hence $\xi_{n-1} = \xi_{n+h-1}$, so that $\{\xi_k\}$ is periodic from the beginning.

5–6 Approximation theorems. Continued fractions provide a very powerful tool in that branch of the theory of numbers known as diophantine approximation. In this subject one is concerned not with equations but with inequalities; the basic problem is to discover what kinds of inequalities involving one or more variables have solutions in integers, or have infinitely many solutions in integers. (The theorems are sometimes phrased in terms of rational numbers, but it comes to the same thing, of course, since rational numbers are quotients of integers.) For example, the fractions (not necessarily reduced) which have fixed denominator q are equally spaced along the real axis, the distance between successive ones being $1/q$:

$$\ldots, -2/q, -1/q, 0/q, 1/q, 2/q, \ldots.$$

Thus every real number x is at most a distance $1/2q$ from such a fraction, and we have the very simple theorem that for every real number x and every integer q, there is an integer p such that $|qx - p| \leq 1/2$. Here we have a diophantine inequality which is solvable for every value of one of the integer variables. Suppose that we asked only that such an inequality be solvable for infinitely many values of q; could we then do better than $1/2$? Is it, for example, possible to find a function $f(q)$ which tends to 0 as q increases without limit, such that the inequality

$$|qx - p| < f(q) \tag{19}$$

has infinitely many integral solutions p, q? The answer to this question is not at all obvious, and it is one of the purposes of the present section to give an answer. The question can easily be refined or generalized in a variety of ways: find the function $f(q)$ which approaches 0 most rapidly and for which (19) has infinitely many solutions for every x, or for every irrational x, or for every x of some other class; or replace the function $qx - p$ by some other function of two variables, or by some function of more than two variables; etc. We shall not attack these more difficult questions here.

The first problem posed above is answered by the following theorem.

THEOREM 5–12. If x is a rational number s/t, then for every rational number p/q different from x, the inequality

$$\left| x - \frac{p}{q} \right| \geq \frac{1}{tq}$$

holds. However, if x is irrational, then there are infinitely many integral solutions p, q of the inequality

$$0 < \left| x - \frac{p}{q} \right| < \frac{1}{q^2}.$$

Thus $f(q) = 1/q$ is an appropriate function in (19), if x is irrational, whereas no function tending to 0 is allowable if x is rational.

Proof: If $x = s/t$ and $p/q \neq s/t$, then

$$\left| x - \frac{p}{q} \right| = \left| \frac{s}{t} - \frac{p}{q} \right| = \frac{|sq - tp|}{tq} \geq \frac{1}{tq},$$

since $|sq - tp|$ is a positive integer. On the other hand, if x is irrational, then it has a nonterminating simple continued fraction, and so infinitely many distinct convergents p_k/q_k, and the desired result follows from (18). ▲

The next theorem shows that the convergents of the continued fraction expansion of x, which we have just seen to be very good approximations to x, are in a certain sense the best possible approximations to x.

THEOREM 5-13. If $n \geq 1$, $0 < q < q_n$ and $p/q \neq p_n/q_n$, then

$$|q_n x - p_n| \leq |qx - p|, \tag{20}$$

with strict inequality unless $n = 1$ and $q_{n+1} = 2$. It follows that under the same circumstances,

$$\left| x - \frac{p_n}{q_n} \right| \leq \left| x - \frac{p}{q} \right|. \tag{21}$$

Proof: Suppose first that $q = q_n$. Then

$$\left| \frac{p_n}{q_n} - \frac{p}{q_n} \right| \geq \frac{1}{q_n},$$

since $p \neq p_n$. On the other hand, by (18),

$$\left| x - \frac{p_n}{q_n} \right| = \frac{1}{q_n(q_n x_{n+1} + q_{n-1})} \leq \frac{1}{q_n q_{n+1}} \leq \frac{1}{2q_n},$$

with strict inequality unless $q_{n+1} = 2$ and $x_{n+1} = 1$, and (21) follows from these two inequalities. When $q = q_n$, (20) and (21) are equivalent, so the theorem is proved in this case. Henceforth we can assume that q is not the denominator of any convergent of x.

Clearly, we can also suppose that $(p, q) = 1$. Let k be the unique index such that $q_{k-1} < q < q_k$; then $1 \leq k \leq n$. If we prove the strict inequality (20) with n replaced by k, then, by (17), it will also hold for n.

Define μ and ν by the equations

$$\mu p_k + \nu p_{k-1} = p,$$

$$\mu q_k + \nu q_{k-1} = q;$$

obviously they are not both negative. Since the determinant of this system is ± 1, it follows from Cramer's rule that μ and ν are the integers

$$\pm \begin{vmatrix} p & p_{k-1} \\ q & q_{k-1} \end{vmatrix} \quad \text{and} \quad \pm \begin{vmatrix} p_k & p \\ q_k & q \end{vmatrix},$$

and neither of these is zero. Since $q = \mu q_k + \nu q_{k-1} < q_k$, the integers

μ and ν must have opposite signs. By Theorem 5–10, the numbers $q_k x - p_k$ and $q_{k-1}x - p_{k-1}$ also have opposite signs. Hence $\mu(q_k x - p_k)$ and $\nu(q_{k-1}x - p_{k-1})$ have the same sign. But

$$qx - p = \mu(q_k x - p_k) + \nu(q_{k-1}x - p_{k-1}),$$

so that

$$|qx - p| > |q_{k-1}x - p_{k-1}| > |q_k x - p_k|. \blacktriangle$$

The next theorem shows that the only solutions of (19) with $f(q) = 1/2q$ are the convergents of x.

THEOREM 5–14. If

$$\left| x - \frac{p}{q} \right| < \frac{1}{2q^2},$$

then p/q is a convergent p_n/q_n of x.

Proof: Suppose that p/q is not a convergent. Then for some index k, $q_{k-1} < q < q_k$, and, as was shown at the end of the proof of the preceding theorem,

$$|q_{k-1}x - p_{k-1}| < |qx - p|.$$

Since $|qx - p| < 1/2q$, we have

$$\left| x - \frac{p_{k-1}}{q_{k-1}} \right| < \frac{1}{2qq_{k-1}}$$

and also

$$\left| x - \frac{p}{q} \right| < \frac{1}{2q^2} < \frac{1}{2qq_{k-1}}.$$

But these two inequalities imply that

$$\left| \frac{p}{q} - \frac{p_{k-1}}{q_{k-1}} \right| < \frac{1}{qq_{k-1}},$$

whereas

$$\left| \frac{p}{q} - \frac{p_{k-1}}{q_{k-1}} \right| = \frac{|pq_{k-1} - qp_{k-1}|}{qq_{k-1}} \geq \frac{1}{qq_{k-1}}.$$

This contradiction shows that p/q is a convergent. ▲

PROBLEMS

1. Why are 22/7 and 355/113 such useful approximations to π?

2. Show, using continued fractions, that if α is a quadratic irrationality, then there is a constant c such that for every pair of integers p and q with $q > 0$,

$$\left| \alpha - \frac{p}{q} \right| > \frac{1}{cq^2}.$$

3. Prove the theorem of Problem 2 without continued fractions. [*Hint:* Let the quadratic equation defining α be $f(x) = ax^2 + bx + c = a(x - \alpha)(x - \alpha') = 0$, where a, b and c are integers. Then $|q^2 f(p/q)|$ is a positive integer, and therefore at least equal to 1.] Generalize the theorem and proof to higher-degree irrationalities.

4. Show that of two consecutive convergents to x, at least one satisfies the inequality

$$\left| x - \frac{p}{q} \right| < \frac{1}{2q^2}.$$

[*Hint:* Show first that

$$\left| \frac{p_{n+1}}{q_{n+1}} - \frac{p_n}{q_n} \right| = \left| x - \frac{p_n}{q_n} \right| + \left| x - \frac{p_{n+1}}{q_{n+1}} \right|,$$

and then give a proof by contradiction.]

5. Below is a sketch of the proof of a theorem. Fill in all details, and state the theorem.

If x is a real number and q is an integer, then the "fractional part" $qx - [qx] = f(q, x)$ satisfies the inequality $0 \leq f(q, x) < 1$. As q takes the values $0, 1, 2, \ldots, n$, there are $n + 1$ points determined in the unit interval, and two of them must lie in some one of the n subintervals

$$0 \leq u < \frac{1}{n}, \frac{1}{n} \leq u < \frac{2}{n}, \ldots, \frac{n-1}{n} \leq u < 1.$$

The distance between these two points is less than $1/n$. Hence . . . .

6. Deduce the second half of Theorem 5–12 from the theorem of the preceding problem.

CHAPTER 6

THE GAUSSIAN INTEGERS

6–1 Introduction. We shall now consider the arithmetical theory of the so-called *Gaussian integers*, these being simply the complex numbers $a + bi$ in which a and b are ordinary integers. To keep matters straight, we shall refer to the numbers $0, \pm 1, \pm 2, \ldots$, which have been the subject of discourse up to now, as the *rational integers*, and designate the set of rational integers by Z. We designate the set of Gaussian integers by $Z[i]$, and use lower-case Greek letters to denote the elements of this set. We sometimes refer to Z and $Z[i]$ as *domains* of integers.

The content of the first four chapters of this book is part of multiplicative number theory, and it all eventually depends on the notion of divisibility. If we attempted to apply similar considerations to the rational numbers or fractions, instead of the rational integers, we should see immediately that everything becomes either trivial or nonsensical. For given two rational numbers A and B, with $B \neq 0$, there is always a rational number C such that $A = BC$ (namely $C = A/B$), and thus every nonzero rational number divides all others. Hence there are no primes and no GCD, every linear equation is solvable, every congruence is true, and so on. In other words, to have an interesting theory, it is necessary to work within a set of numbers in which division is not always possible (that is, not all quotients of elements belong to the set). On the other hand, it simplifies matters greatly if the usual rules of algebra concerning multiplication, addition, and subtraction, as embodied in postulates I through IV of Chapter 1, continue to hold. The Gaussian integers meet both of these requirements, and have, in addition, a number of special properties which make it possible to build up a theory remarkably similar to that already developed for rational integers. The bulk of the present chapter will be devoted to this simple case, but in the final section we shall consider the more complicated situation in which the integers are taken to be the elements $a + b\sqrt{10}$ of $Z[\sqrt{10}]$.

These generalizations of "rational" number theory should not be regarded as mere curiosities. They are special instances of a much broader development, called the theory of algebraic numbers, in which one considers general algebraic irrationalities—roots of algebraic equations of all degrees —rather than just certain quadratic irrationalities. This theory, which is deep and difficult, is not only interesting in its own right, but it has many applications in rational number theory, and its more comprehensive viewpoint makes possible a real understanding of various phenomena in rational number theory which would otherwise remain completely mysterious.

6-2 Divisibility, units, and primes. Let α and β be Gaussian integers, with $\beta \neq 0$. We say that β *divides* α, and write $\beta|\alpha$, if there is an element γ of $Z[i]$ such that $\alpha = \beta\gamma$. For example, $(1+i)|2$ since $2 = (1+i)(1-i)$, and also $(1+i)|(1-i)$ since $1+i = (1-i)\cdot i$. On the other hand, $(1+i)\nmid(1+2i)$, for supposing the opposite leads to the equation

$$1 + 2i = (1+i)(a+bi) = (a-b) + (a+b)i,$$

from which it follows, by comparison of real and imaginary parts, that $a - b = 1$ and $a + b = 2$. Adding these equations, we obtain $2a = 3$, which is false for every a in Z.

We must verify that this definition of divisibility is consistent with that given earlier for rational integers, for otherwise we should have to use different symbols to indicate the domains with respect to which divisibility is asserted. That is, it is conceivable that $2|7$ when 2 and 7 are regarded as elements of $Z[i]$, because there might be a Gaussian integer α such that $7 = 2\alpha$, even though there is no rational integer a such that $7 = 2a$. In fact, such a thing never happens. For if a and $b \neq 0$ are in Z, and $b|a$ in $Z[i]$, then for some $\gamma = c + di$ we have $a = b\gamma$. But the equation

$$a = b\gamma = bc + bdi$$

yields $a = bc$ and $bd = 0$; thus $d = 0$, γ is real and is therefore in Z, and $b|a$ in $Z[i]$ implies $b|a$ in Z.

If $\alpha = a + bi$, then the nonnegative rational integer

$$(a+bi)(a-bi) = a^2 + b^2$$

is called the *norm* of α, and is designated by $\mathbf{N}\alpha$ or $\mathbf{N}(\alpha)$. By writing out the multiplication, it is easily verified that for every α and β in $Z[i]$,

$$\mathbf{N}(\alpha\beta) = \mathbf{N}\alpha\mathbf{N}\beta. \tag{1}$$

We express this by saying that the norm is *multiplicative.* It follows from (1) that if $\alpha|\gamma$, then $\mathbf{N}\alpha|\mathbf{N}\gamma$.

Certain Gaussian integers divide every Gaussian integer; these are called the *units* of $Z[i]$. In particular, a unit must divide 1. Conversely, if $\epsilon|1$, then ϵ is a unit, for we then have $1 = \epsilon\eta$, and so for every α in $Z[i]$ we can write $\alpha = \alpha \cdot 1 = (\alpha\eta)\epsilon$, whence $\epsilon|\alpha$. So we can find the units of $Z[i]$ simply by finding the divisors of 1. Now if $\epsilon|1$, then $\mathbf{N}\epsilon|\mathbf{N}1$, so $\mathbf{N}\epsilon = 1$. The only solutions of the equation $a^2 + b^2 = 1$ in Z are $\pm 1, 0$ and $0, \pm 1$, and hence the only units of $Z[i]$ are ± 1 and $\pm i$.

If ϵ is a unit and α and β are elements of $Z[i]$ such that $\alpha = \beta\epsilon$, then α and β are said to be *associates.* Note that under this definition, α is an associate of itself.

A Gaussian integer which has no other divisors than its associates and the units is said to be *prime in* $Z[i]$ or, if the context is well-understood, it is simply said to be prime. This point has to be emphasized since in the present context the rational integers do behave differently when considered as elements of Z or of $Z[i]$, some of them being prime in the first domain and not in the second. For as we saw above, the rational prime 2 splits in a nontrivial way, and is therefore not prime, in $Z[i]$:

$$2 = i(1 - i)^2.$$

(Here the factor $1 - i$ is prime, because if $1 - i = \alpha\beta$, then $2 = \mathbf{N}\alpha\mathbf{N}\beta$, and so either $\mathbf{N}\alpha = 1$ or $\mathbf{N}\beta = 1$.) On the other hand, not every rational prime splits in $Z[i]$; for example, 3 does not. For if $\alpha | 3$, then $\mathbf{N}\alpha | 9$, so that if α is neither a unit (with norm 1) nor an associate of 3 (with norm 9), it must be that $\mathbf{N}\alpha = 3$. But the equation $a^2 + b^2 = 3$ has no solution in Z.

Sometimes we shall distinguish the two notions of primality by referring to Gaussian primes and rational primes.

We have so far only two examples of Gaussian primes: $1 - i$ (or its associates $\pm 1 \pm i$) and 3. We pause to show that, in fact, there are infinitely many primes in $Z[i]$.

THEOREM 6–1. There are infinitely many rational primes of the form $4k - 1$.

Proof: Every odd prime is congruent either to 1 or to -1 (mod 4). Hence every odd number is congruent (mod 4) to 1 or -1, according as the number of its prime factors of the form $4k - 1$ is even or odd. In particular, if $n \equiv -1 \pmod 4$, then n contains an odd number of prime factors, and therefore at least one, of the form $4k - 1$.

Now let p_k be the kth prime, and set $N = 4p_1p_2 \cdots p_n - 1$. Every prime divisor of N is different from any of $p_1, p_2, \ldots, p_n$, and hence is larger than p_n. By the preceding paragraph, N has a prime factor of the form $4k - 1$. We have therefore shown that for every n, there is a prime of the form $4k - 1$ which is larger than p_n. This implies the theorem. ▲

THEOREM 6–2. There are infinitely many Gaussian primes.

Proof: We prove this by proving a stronger statement, namely that infinitely many rational primes are also Gaussian primes. In fact, we shall show that every rational prime $p \equiv -1 \pmod 4$ is also a Gaussian prime. Suppose that $p = \alpha\beta$, so that $\mathbf{N}p = p^2 = \mathbf{N}\alpha\mathbf{N}\beta$. If $\mathbf{N}\alpha = 1$, then α is a unit. If $\mathbf{N}\alpha = p^2$, then $\mathbf{N}\beta = 1$, so β is a unit. Hence the supposed factorization is trivial unless $\mathbf{N}\alpha = \mathbf{N}\beta = p$. However, if $\alpha = a + bi$, we then have

$$a^2 + b^2 = p,$$

$$a^2 + b^2 \equiv 3 \pmod 4.$$

But a square is either 0 or 1 (mod 4), and no two numbers, each of which is 0 or 1, add to 3. ▲

We shall see later that the rational primes of the form $4k + 1$ always split as the product of two nonassociated Gaussian primes, and that this exhausts the set of Gaussian primes. Thus every Gaussian prime is a factor of exactly one rational prime.

Problems

1. Show that $\mathbf{N}\alpha | \mathbf{N}\beta$ does not imply $\alpha | \beta$.

2. Show that associates have the same norm, but that two Gaussian integers having the same norm need not be associates.

3. Show that if $\mathbf{N}\alpha$ is a rational prime, than α is a Gaussian prime.

4. Show that $(1 + i) \nmid (1 + 2i)$ by direct consideration of the fraction $(1 + 2i)/(1 + i)$

5. Show that $(1 + i) \nmid (1 + 2i)$ using the multiplicativity of the norm.

6. Show that a Gaussian integer has only finitely many divisors in $Z[i]$. Find all the divisors of 10, and prove that there are no others. Do the same for $3 + 7i$.

7. Use the multiplicativity of the norm to show that if each of two rational integers m and n is the sum of two rational squares, the same is true of the product mn.

6–3 The greatest common divisor. Up to this point, nothing has been said about inequalities in $Z[i]$. As a matter of fact, there is no way to introduce the relation "$<$" in $Z[i]$ in such a way that the following two statements hold:

(a) For any two elements α and β of $Z[i]$, exactly one of the relations $\alpha < \beta$, $\alpha = \beta$, or $\alpha > \beta$ holds.

(b) If $\alpha < \beta$ and $0 < \gamma$, then $\alpha\gamma < \beta\gamma$.

To see this, we note first that under any definition of "$\alpha < \beta$" which is consistent with (a) and (b), necessarily $0 < 1$. For if not, then by (a) it must be that $0 < -1$. But then by (b), with $\alpha = 0$ and $\beta = \gamma = -1$, we have $0 < (-1)^2 = 1$, a contradiction.

We complete the proof of the asserted impossibility by showing that neither of the relations $i < 0$ or $0 < i$ can hold. For if $i < 0$, then $0 < -i$, and hence $0 < (-i)^2 = -1$, which is false. If $0 < i$, then $0 < i^2 = -1$, which is also false.

Since the entire theory of rational integers developed earlier depended ultimately on Theorem 1–1, which involves an inequality, it is important to note that a weak sort of comparison of elements of $Z[i]$ can be effected by comparing norms. This is what is done in the following analog of Theorem 1–1.

THEOREM 6–3. If α and β are integers of $Z[i]$, and $\beta \neq 0$, then there are κ and ρ in $Z[i]$ such that

$$\alpha = \beta\kappa + \rho, \qquad \mathbf{N}\rho < \mathbf{N}\beta.$$

Proof: Since $\beta \neq 0$, we can write

$$\frac{\alpha}{\beta} = \frac{a + bi}{c + di} = \frac{(a + bi)(c - di)}{c^2 + d^2} = A + Bi,$$

where A and B are rational numbers, not necessarily integers. Let x and y be the rational integers nearest to A and B, respectively, so that

$$|A - x| \leq \tfrac{1}{2},$$
$$|B - y| \leq \tfrac{1}{2}.$$

Then

$$\left|\frac{\alpha}{\beta} - (x + yi)\right| = |(A - x) + (B - y)i|$$
$$= ((A - x)^2 + (B - y)^2)^{1/2} \leq (\tfrac{1}{4} + \tfrac{1}{4})^{1/2} < 1.$$

Hence, if we set

$$x + yi = \kappa, \qquad \alpha - \beta(x + yi) = \rho,$$

then

$$\mathbf{N}\rho = \mathbf{N}(\alpha - \kappa\beta) = \mathbf{N}\beta \cdot \mathbf{N}\left(\frac{\alpha}{\beta} - \kappa\right) < \mathbf{N}\beta,$$

and κ and ρ are in $Z[i]$. ▲

On the basis of Theorem 6–3, the Euclidean algorithm can now be generalized to Gaussian integers, as follows:

$$\alpha = \beta\kappa_1 + \rho_1, \qquad \mathbf{N}\rho_1 < \mathbf{N}\beta,$$
$$\beta = \rho_1\kappa_2 + \rho_2, \qquad \mathbf{N}\rho_2 < \mathbf{N}\rho_1,$$
$$\vdots$$
$$\rho_{k-2} = \rho_{k-1}\kappa_k + \rho_k, \qquad \mathbf{N}\rho_k < \mathbf{N}\rho_{k-1},$$
$$\rho_{k-1} = \rho_k\kappa_{k+1}.$$

The sequence of equations must terminate, because $\mathbf{N}\beta$, $\mathbf{N}\rho_1$, $\mathbf{N}\rho_2, \ldots$ is a decreasing sequence of nonnegative integers. It can be shown that ρ_k, the last nonvanishing remainder, is a divisor of both α and β, by working up from the last equation to the first, and it can be shown that every common divisor of α and β is a divisor of ρ_k, by working down from the first equation to the last, just as was done in Section 2–2. From the next to last equation, ρ_k can be written as a linear combination of ρ_{k-1} and

ρ_{k-2}, and then $\rho_{k-1}, \rho_{k-2}, \ldots, \rho_1$ can be successively eliminated with the help of the earlier equations to yield ρ_k as a linear combination of α and β. Thus ρ_k is a Gaussian integer having all the properties of the number δ listed in the following theorem.

THEOREM 6-4. Let α and β be Gaussian integers, not both 0. Then there is an integer δ of $Z[i]$ with the following properties:

(i) $\delta | \alpha$ and $\delta | \beta$.
(ii) If δ' is any integer such that $\delta' | \alpha$ and $\delta' | \beta$, then $\delta' | \delta$.
(iii) There are ξ and η in $Z[i]$ such that $\delta = \alpha\xi + \beta\eta$.

Any two integers δ_1 and δ_2 having properties (i) and (ii) are associates.

Proof: It is only the uniqueness of δ (except for a possible unit factor) which remains in doubt. Suppose that δ_1 and δ_2 are two Gaussian integers having properties (i) and (ii). Then since $\delta_1 | \alpha$ and $\delta_1 | \beta$, it follows from (ii), with $\delta' = \delta_1$ and $\delta = \delta_2$, that $\delta_1 | \delta_2$. By symmetry, we also have $\delta_2 | \delta_1$. Hence δ_1 and δ_2 are associates. ▲

Any Gaussian integer δ having properties (i) and (ii) is called a GCD of α and β, and we write $(\alpha, \beta) = \delta$. [Strictly speaking, we should of course write "$(\alpha, \beta) = \pm\delta$ or $\pm i\delta$."] If $(\alpha, \beta) = 1$, we say that α and β are relatively prime.

As an example, suppose we wish to find a GCD of $7 + 11i$ and $3 + 5i$. We have

$$\frac{7 + 11i}{3 + 5i} = \frac{76 - 2i}{34} = 2 + \frac{8 - 2i}{34},$$

so that

$$7 + 11i = (3 + 5i)2 + \frac{8 - 2i}{3 - 5i} = (3 + 5i)2 + (1 + i).$$

Similarly,

$$\frac{3 + 5i}{1 + i} = \frac{8 + 2i}{2} = 4 + i,$$

$$3 + 5i = (4 + i)(1 + i).$$

Thus $(7 + 11i, 3 + 5i) = 1 + i$.

PROBLEMS

1. Compute the following GCD's:
(a) $(16 - 2i, 33 + 17i)$, (b) $(4 + 6i, 7 - i)$, (c) $(5 + i, 4 - 3i)$.

2. Find the exact conditions under which $(1 + i) | (a + bi)$.

3. Express each of the GCD's in Problem 1 as a linear combination of the two entries, with coefficients in $Z[i]$.

4. Show that $(\mu\alpha, \mu\beta) = \mu(\alpha, \beta)$ if $\mu \neq 0$.

5. Show that if $\mu|\alpha$ and $\mu|\beta$, then $(\alpha/\mu, \beta/\mu) = (\alpha, \beta)/\mu$.

6. Show that if α is relatively prime to each of the numbers $\beta_1, \ldots, \beta_n$, then α is relatively prime to $\beta_1 \cdots \beta_n$.

7. Show that if $\beta|\alpha$, $\gamma|\alpha$, and $(\beta, \gamma) = 1$, then $\beta\gamma|\alpha$.

8. Find an analog of Theorem 6–3 in the domain $Z[\sqrt{-2}]$ of numbers of the form $a + b\sqrt{-2}$, with a and b in Z. Is Theorem 6–4 valid in this domain? What happens in the domain $Z[\sqrt{-5}]$?

6–4 The unique factorization theorem in $Z[i]$. The next four theorems are analogs of the theorems of Section 2–3. The only significant change is that inequalities between integers have been replaced by inequalities between their norms.

THEOREM 6–5. Every integer α with $\mathbf{N}\alpha > 1$ can be represented as a finite product of primes.

Proof: The proof is carried out by induction on $\mathbf{N}\alpha$. The smallest case to consider is $\mathbf{N}\alpha = 2$: the four Gaussian integers of norm 2 are associates of the prime $1 + i$, and are therefore primes themselves. Suppose then that α is an integer with the property that every integer of smaller norm (and not a unit) has a representation of the required sort. If α is itself prime, we are through. Otherwise α has a decomposition $\alpha = \beta\gamma$, with $1 < \mathbf{N}\beta < \mathbf{N}\alpha$ and $1 < \mathbf{N}\gamma < \mathbf{N}\alpha$. The induction hypothesis then implies that

$$\beta = \pi_1'\pi_2' \cdots \pi_s' \quad \text{and} \quad \gamma = \pi_1''\pi_2'' \cdots \pi_t'',$$

where the π_i' and π_i'' are primes in $Z[i]$, and hence

$$\alpha = \pi_1' \cdots \pi_s'\pi_1'' \cdots \pi_t''. \blacktriangle$$

THEOREM 6–6. If $\alpha|\beta\gamma$ and $(\alpha, \beta) = 1$, then $\alpha|\gamma$.

Proof: If $(\alpha, \beta) = 1$, there are integers ξ and η such that $\alpha\xi + \beta\eta = 1$, and therefore $\alpha\gamma\xi + \beta\gamma\eta = \gamma$. But α divides both $\alpha\gamma$ and $\beta\gamma$, and hence the left side of the last equation, and therefore α divides γ. ▲

THEOREM 6–7. If $\pi, \pi_1, \ldots, \pi_n$ are Gaussian primes, and $\pi|\pi_1 \cdots \pi_n$, then for at least one m, π is an associate of π_m.

Proof: Suppose that $\pi|\pi_1 \cdots \pi_n$, but that π is different from any of $\pi_1, \ldots, \pi_{n-1}$ or their associates. Then π is relatively prime to each of $\pi_1, \ldots, \pi_{n-1}$, and so is relatively prime to the product $\pi_1 \cdots \pi_{n-1}$. (This follows from the fact that if $(\alpha, \beta) = 1$ and $(\alpha, \gamma) = 1$, then $(\alpha, \beta\gamma) = 1$, an implication which in turn follows by multiplying together the equations $\alpha\xi + \beta\eta = 1$ and $\alpha\mu + \gamma\nu = 1$ to obtain 1 as the linear combination $1 = \alpha(\alpha\xi\mu + \beta\eta\mu + \gamma\nu\xi) + \beta\gamma \cdot \eta\nu$ of α and $\beta\gamma$.) By Theorem 6-6, $\pi | \pi_n$, so that π and π_n are associates. ▲

THEOREM 6-8 (*Unique Factorization Theorem for* $Z[i]$). The representation of each Gaussian integer α with $\mathbf{N}\alpha > 1$ as a product of primes is unique except for the order of factors and the presence of units.

Proof: Suppose that α is any element of $Z[i]$ with $\mathbf{N}\alpha > 1$ and having the two factorizations

$$\alpha = \pi_1 \cdots \pi_r = \pi_1' \cdots \pi_s'.$$

Then $\pi_1 | \pi_1' \cdots \pi_s'$, so by Theorem 6-7, π_1 is an associate of one of the π_i', which we may as well take to be π_1'. Cancelling π_1 from the above equation, we obtain

$$\frac{\alpha}{\pi_1} = \pi_2 \cdots \pi_r = \epsilon_1 \pi_2' \cdots \pi_s',$$

where ϵ_1 is a unit. The argument can now be repeated, with the primes from the two factorizations successively paired off and cancelled, there being equality to within a unit factor between the remaining prime products at each stage. When all primes $\pi_1, \ldots, \pi_r$ have been eliminated, all primes must be gone from the other factorization too, with just a unit remaining, and this unit must in fact be 1. ▲

PROBLEMS

1. Show that if α is not prime, it has a prime factor π with $\mathbf{N}\pi \leq \sqrt{\mathbf{N}\alpha}$.

2. (a) List all Gaussian integers $\alpha = a + bi$ with $\mathbf{N}\alpha \leq 9$ which lie in the first quadrant, that is, those for which $a > 0$ and $b \geq 0$. Multiply the numbers in this list by $1 + i$, and find the associates in the first quadrant of these multiples. Using Problem 1, find all Gaussian primes π with $\mathbf{N}\pi \leq 9$.

(b) Continuing, list all Gaussian integers α in the first quadrant with $\mathbf{N}\alpha \leq 81$, and then find all primes π with $\mathbf{N}\pi \leq 81$.

3. Find the prime decompositions of (a) $5 + 6i$, (b) $7 - 3i$.

6-5 The primes in $Z[i]$. We have already seen that in $Z[i]$, the rational prime 2 splits as the product of the associated primes $1 + i$ and $1 - i$, and that the rational primes $p \equiv 3 \pmod 4$ remain prime. All other rational primes are of the form $4k + 1$, and we now turn our attention to these.

THEOREM 6-9. Every rational prime $p \equiv 1 \pmod 4$ splits in $Z[i]$ as the product of two nonassociated Gaussian primes.

Proof: Suppose that $p \equiv 1 \pmod 4$. Then according to Euler's criterion (Theorem 3–15), -1 is a quadratic residue of p, and hence there is a rational integer x such that $x^2 \equiv -1 \pmod p$. Thus $p|(x^2+1)$. Now in $Z[i]$ we have

$$x^2 + 1 = (x - i)(x + i),$$

and if p were prime in $Z[i]$, then it would have to divide one of the factors $x - i$ and $x + i$. But this is obviously not the case, since $p \nmid \pm 1$. Therefore p splits in $Z[i]$ into two non-unit factors.

Suppose that there is a factorization

$$p = \alpha\beta, \tag{2}$$

where we do not suppose that α and β are prime, but only that they are not units. Then $\mathbf{N}p = p^2 = \mathbf{N}\alpha\mathbf{N}\beta$, and since $\mathbf{N}\alpha$ and $\mathbf{N}\beta$ are rational integers different from 1, it must be that

$$\mathbf{N}\alpha = \mathbf{N}\beta = p. \tag{3}$$

If α were not prime, we should have

$$\alpha = \pi_1\pi_2 \cdots \pi_r, \qquad r > 1,$$

and hence

$$p = \mathbf{N}\pi_1\mathbf{N}\pi_2 \cdots \mathbf{N}\pi_r,$$

which is impossible since p is a rational prime. Similarly, β must be prime.

From (2) and (3) we have three decompositions for p,

$$p = \alpha\beta = \alpha\bar{\alpha} = \beta\bar{\beta},$$

and from the first two we see that $\beta = \bar{\alpha}$. Thus $\bar{\alpha}$ is also prime, and p is the product of two complex-conjugate Gaussian primes:

$$p = \pi\bar{\pi}. \tag{4}$$

It remains to be shown that these primes are not associated. Let

$$\pi = a + bi,$$

and assume that $\bar{\pi} = \epsilon\pi$ for some unit ϵ. For the four units, we make the following deductions by comparing real and imaginary parts in the equation $a - bi = \epsilon(a + bi)$:

$$\begin{aligned} \epsilon &= 1: & b &= 0, \\ \epsilon &= -1: & a &= 0, \\ \epsilon &= i: & a &= -b, \\ \epsilon &= -i: & a &= b. \end{aligned}$$

But none of these alternatives is possible, because of (4):

$$\begin{aligned} b = 0 &\quad \text{implies} \quad p = a^2, \\ a = 0 &\quad \text{implies} \quad p = b^2, \\ a = -b &\quad \text{implies} \quad p = a^2(1 - i)(1 + i) = 2a^2, \\ a = b &\quad \text{implies} \quad p = a^2(1 + i)(1 - i) = 2a^2. \end{aligned}$$

Hence $\bar{\pi} \not\asymp \epsilon\pi$. ▲

Equation (4) shows that every $p \equiv 1 \pmod 4$ is the norm of a Gaussian prime: $p = (a + bi)(a - bi) = a^2 + b^2$. This has the following immediate consequence in rational number theory:

THEOREM 6–10. Every prime of the form $4k + 1$ can be represented as the sum of two squares, $p = a^2 + b^2$, and this representation is unique except for order and sign of a and b.

This theorem was discovered empirically by Fermat, and proved by Euler. There are proofs which are not based on Gaussian integers, but the present proof indicates clearly how the theory of other kinds of integers can be used to obtain information about the rational integers.

We now show that we have found all the primes in $Z[i]$.

THEOREM 6–11. The associates of the following represent all the Gaussian primes:

the associated divisors $1 + i$ and $1 - i$ of 2,
the rational primes $p \equiv 3 \pmod 4$,
the nonassociated conjugate prime divisors $a + bi$ and $a - bi$ of the rational primes $p \equiv 1 \pmod 4$.

Proof: Let π be a Gaussian prime, and let $\mathbf{N}\pi = \pi\bar{\pi} = a$. The rational integer a can have at most two rational prime factors, since otherwise it would have more than two prime factors in $Z[i]$, which is not the case. Moreover, if $a = pq$, where p and q are rational primes, then $p = q$. For suppose that $p \neq q$. Then $|p| \neq |q|$, whereas $|\pi| = |\bar{\pi}|$, and the unique factorization theorem for $Z[i]$ is violated, whether or not p and q are also prime in $Z[i]$.

Thus either $\mathbf{N}\pi = p^2$ or $\mathbf{N}\pi = p$, where p is a rational prime. In the first case, $\pi\bar{\pi} = p \cdot p$, so $\pi = \bar{\pi} = p$, and as we know, this happens exactly when $p \equiv 3 \pmod 4$. On the other hand, if $\mathbf{N}\pi = p$, then π and $\bar{\pi}$ are the nonreal Gaussian prime factors of p, and there are such factors when $p = 2$ and when $p \equiv 1 \pmod 4$. ▲

PROBLEMS

1. Using the theorems of this section, list all Gaussian primes π with $\mathbf{N}\pi \leq 100$.

2. Find the Gaussian prime decompositions of (a) $7 + 6i$, (b) $7 + 5i$, (c) $8 + 5i$. [*Hint:* First factor $\mathbf{N}\alpha$ in Z.]

3. (a) Show that if $p \equiv 3 \pmod 4$ is a rational prime and n is an integer, then $p \nmid (n^2 + 1)$. (Use the argument of the first paragraph of the proof of Theorem 6–9.)

(b) Let 5, 13, 17, ..., p_k be the first k primes $p \equiv 1 \pmod 4$. By considering the rational prime factors of $N = (5 \cdot 13 \cdot 17 \cdots p_k)^2 + 1$, show that there are infinitely many $p \equiv 1 \pmod 4$, and hence that there are infinitely many nonreal Gaussian primes.

6–6 Another quadratic domain. For purposes of contrast, we terminate the discussion of quadratic arithmetic with a brief description of a domain of integers in which matters are not so simple as in Z and $Z[i]$. This domain is the set, which we call $Z[\sqrt{10}]$, of numbers of the form $a + b\sqrt{10}$, where a and b are rational integers. It is clear that the sum, difference, and product of elements of $Z[\sqrt{10}]$ are again in $Z[\sqrt{10}]$. We shall use capital letters $A, B, \ldots$ to designate elements of $Z[\sqrt{10}]$.

Divisibility, units, and primes can be defined exactly as before:

We say that B *divides* A, and write $B|A$, if there is a C such that $A = BC$.

An element E of $Z[\sqrt{10}]$ is called a *unit* if $E|1$.

An element P of $Z[\sqrt{10}]$ is said to be *prime* in $Z[\sqrt{10}]$ if in every factorization $P = AB$, either A or B is a unit.

The *norm* $\mathbf{N}A$ of the integer $A = a + b\sqrt{10}$ is the product $(a + b\sqrt{10}) \times (a - b\sqrt{10}) = a^2 - 10b^2$. It is easily seen to be multiplicative, so that $\mathbf{N}(AB) = \mathbf{N}A\mathbf{N}B$. This implies that the units are exactly the integers with norm ± 1:

$$1 = EF,$$
$$\mathbf{N}1 = 1 = \mathbf{N}E\mathbf{N}F,$$
$$\mathbf{N}E = \mathbf{N}F = \pm 1.$$

It is no longer the case that the norm is always nonnegative, so we must allow the possibility of $\mathbf{N}E = -1$. A more serious complication is that there are now infinitely many units. For the equation $\mathbf{N}E = a^2 - 10b^2 = -1$ has the solution $a = 3$, $b = 1$, so $E = 3 + \sqrt{10}$ is a unit. Since $\mathbf{N}(E^n) = (\mathbf{N}E)^n = (-1)^n$, every power of E is also a unit. Since $E > 1$, it is clear that $\cdots < E^{-1} < 1 < E < E^2 < E^3 < \cdots$, and hence these powers of E are distinct, and they constitute an infinite set of units. It should be noted that the units, with the exception of ± 1, do not have absolute value 1, so that associates (elements differing only by a unit factor) have the same "size" only in the sense of norm, and not in absolute value. It is fortuitous that norm and absolute value nearly coincide for Gaussian integers.

Another difficulty as regards $Z[\sqrt{10}]$ is the absence of a unique factorization theorem. For example, the element 6 of $Z[\sqrt{10}]$ has two genuinely different prime factorizations, namely

$$6 = 2 \cdot 3 = (4 + \sqrt{10})(4 - \sqrt{10}).$$

To see that these really are different, it suffices to show that 2, 3, and $4 + \sqrt{10}$ are prime in $Z[\sqrt{10}]$, and that neither $4 + \sqrt{10}$ nor $4 - \sqrt{10}$ is an associate of 2.

To prove the primality of 2, 3, and $4 + \sqrt{10}$, we note first that for no A in $Z[\sqrt{10}]$ do we have $|\mathbf{N}A| = 2$ or 3. For the congruences $a^2 \equiv \pm 2$ or $\pm 3 \pmod{10}$ are insolvable (in other language, the decimal expansion of the square of a rational integer never terminates in 2, 3, 7, or 8), and hence the equations $a^2 - 10m = \pm 2$ or ± 3 are insolvable in Z, from which it follows that the equations $a^2 - 10b^2 = \pm 2$ or ± 3 also are insolvable in Z.

If $2 = AB$, then $4 = \mathbf{N}A\mathbf{N}B$, and since $|\mathbf{N}A| \neq 2$, either $|\mathbf{N}A|$ or $|\mathbf{N}B|$ is 1. Thus 2 is prime in $Z[\sqrt{10}]$. If $3 = AB$, then $9 = \mathbf{N}A\mathbf{N}B$, and since $|\mathbf{N}A| \neq 3$, either $|\mathbf{N}A|$ or $|\mathbf{N}B|$ is 1. Thus 3 is prime in $Z[\sqrt{10}]$. Finally, if $4 + \sqrt{10} = AB$, then $6 = \mathbf{N}A\mathbf{N}B$, and since $|\mathbf{N}A| \neq 2, 3$, either $|\mathbf{N}A|$ or $|\mathbf{N}B|$ is 1. Thus $4 + \sqrt{10}$ is prime in $Z[\sqrt{10}]$.

To see that neither $4 + \sqrt{10}$ nor $4 - \sqrt{10}$ is an associate of 2, we compare norms. If $2 = E(4 + \sqrt{10})$, where E is a unit of $Z[\sqrt{10}]$, then $\mathbf{N}2 = 4 = \mathbf{N}E \cdot \mathbf{N}(4 + \sqrt{10}) = \pm 1 \cdot 6$, which is false. Similarly, $2 \neq E(4 - \sqrt{10})$. ▲

This example shows that there is no Euclidean algorithm in $Z[\sqrt{10}]$, and hence not even a division theorem of the customary sort. That is, it may not be possible to find Q and R such that $A = BQ + R$, if it is required that $|\mathbf{N}R|$ be smaller than $|\mathbf{N}B|$. Moreover, two integers of $Z[\sqrt{10}]$ do not always have a GCD which is a linear combination of those integers. It should be apparent that the lack of unique factorization in $Z[\sqrt{10}]$ is a much more serious difficulty than the existence of infinitely many units; the latter is surprising at first, and leads to some slight complications, but it does not invalidate whole sections of the usual arithmetic theory. The attempt to restore unique factorization in domains such as $Z[\sqrt{10}]$ was one of the starting points of modern abstract algebra. The successful solution of the problem is too lengthy for inclusion here.

CHAPTER 7

DIOPHANTINE EQUATIONS

7–1 Introduction. As was mentioned in Chapter 1, there is rather little systematic knowledge that could be called a general theory of Diophantine equations, especially of equations of degree larger than 2. Sometimes a method especially devised for one problem has applications elsewhere, of course; this is true, for instance, of the so-called method of infinite descent invented by Fermat, of which we shall give an example when we consider the equation $x^4 + y^4 = z^4$. (This is really an inductive proof in disguise; an equation is shown to have no solution by supposing it has solutions, considering a "smallest" solution in some sense, and deriving a still smaller solution.) But all too frequently the proofs are entirely *ad hoc*, and of no use for new problems.

We shall consider here, in addition to the quartic equation above, the Pythagorean equation $x^2 + y^2 = z^2$ and the Pell equation $x^2 - dy^2 = 1$. Both of these equations have infinitely many solutions, which we shall be able to describe completely, but in quite different ways.

7–2 The equation $x^2 + y^2 = z^2$. If we know all the *primitive* solutions of this equation, in which $(x, y, z) = 1$, then we can find all other solutions by multiplying x, y, and z by an arbitrary integer d. Among the primitive solutions it suffices to find those for which x, y, and z are positive, since all others arise from positive solutions by simple sign changes. Finally, in any primitive solution exactly one of x and y must be odd, for at least one must be odd to give a primitive solution, and if both were odd we should have $x^2 + y^2 \equiv 2 \pmod 4$, while $z^2 \equiv 1$ or $0 \pmod 4$. We shall discuss the solutions in which x is odd; because of the symmetry of the equation in x and y, the solutions in which y is odd can be obtained simply by permuting x and y in the solutions now to be described.

THEOREM 7–1. A general primitive solution of

$$x^2 + y^2 = z^2, \qquad y \text{ even}, \qquad x > 0, \quad y > 0, \quad z > 0,$$

is given by

$$x = a^2 - b^2, \qquad y = 2ab, \qquad z = a^2 + b^2,$$

where a and b are prime to each other and not both odd, and $a > b > 0$.

Proof: It is easily verified that for every such pair of integers a and b, the corresponding integers x, y, and z satisfy all the requirements. It remains to show that every solution arises from suitably chosen a and b satisfying the conditions of the theorem.

Suppose that $x^2 + y^2 = z^2$. Since $(x, y, z) = 1$, we also have $(y, z) = 1$, so that $(z - y, z + y) = 1$ or 2. But z is odd and y is even, and therefore $(z - y, z + y) = 1$. Hence from the equation

$$x^2 = (z - y)(z + y),$$

we deduce that $z - y$ and $z + y$ must be odd squares, since they are positive. Now if t and u are integers of the same parity (both even or both odd), there are integers a and b such that $t = a + b$ and $u = a - b$, namely $a = (t + u)/2$ and $b = (t - u)/2$. Applying this in the case where t and u are the odd numbers of which $z + y$ and $z - y$ are the squares, respectively, we can set

$$z - y = (a - b)^2, \qquad z + y = (a + b)^2,$$

whence

$$z = \frac{(a - b)^2 + (a + b)^2}{2} = a^2 + b^2,$$

$$y = \frac{(a + b)^2 - (a - b)^2}{2} = 2ab,$$

$$x = (a - b)(a + b) = a^2 - b^2.$$

Since $(z - x, z + x) = 2$ because z and x are relatively prime and both odd, and since $z - x = 2a^2$ and $z + x = 2b^2$, it must be that $(a, b) = 1$. Since x is odd, $a + b$ must be odd. Since $y > 0$, a and b must have the same sign, and since x is positive, $|a| > |b|$. Finally, since the pairs a, b and $-a, -b$ yield the same solution, we can suppose that $a > b > 0$. ▲

Problems

1. Referring to Theorem 7–1, show that every solution x, y arises from just one pair a, b fulfilling the requirements mentioned there.

2. Let p be a prime, and suppose that $x^2 + py^2 = z^2$, where $(x, y, z) = 1$. Show that, except for the signs of x, y, and z, either

$$x = \frac{u^2 - pv^2}{2}, \qquad y = uv, \qquad z = \frac{u^2 + pv^2}{2}, \qquad u \text{ and } v \text{ both odd},$$

or

$$x = u^2 - pv^2, \qquad y = 2uv, \qquad z = u^2 + pv^2, \qquad \text{exactly one of } u \text{ and } v \text{ odd}.$$

7–3 The equation $x^4 + y^4 = z^4$. According to Fermat's conjecture, the equation $x^n + y^n = z^n$ never has a solution in nonzero integers x, y, z if $n > 2$. Various necessary conditions for the existence of a solution are known, and from these it is possible to show that there is no solution for many different values of n, but the general conjecture has been neither proved nor disproved. Indeed, it is not even known whether there can be infinitely many solutions, for certain $n > 2$.

If $n > 2$, then n is divisible either by 4 or by some odd prime, and we call this divisor r, whichever it may be. Then $n = rm$ for suitable m, and the equation $x^n + y^n = z^n$ is the same as $(x^m)^r + (y^m)^r = (z^m)^r$. Hence, if it could be shown that the equation $X^r + Y^r = Z^r$ has no nonzero solution, then, in particular, there would be no solution $X = x^m$, $Y = y^m$, $Z = z^m$, and consequently no solution of $x^n + y^n = z^n$. Thus it suffices to consider the Fermat equation for $n = 4$ or an odd prime. We now treat the case $n = 4$.

THEOREM 7–2. The equation $x^4 + y^4 = z^4$ is not solvable in nonzero integers.

Proof: It suffices to show that there is no primitive solution of the equation

$$x^4 + y^4 = z^2.$$

Suppose that x, y, and z constitute such a solution; with no loss in generality we may take $x > 0$, $y > 0$, $z > 0$, and y even. Writing the supposed relation in the form

$$(x^2)^2 + (y^2)^2 = z^2,$$

we see from Theorem 7–1 that

$$x^2 = a^2 - b^2, \qquad y^2 = 2ab, \qquad z = a^2 + b^2,$$

where $(a, b) = 1$ and exactly one of a and b is odd. If a were even, we would have

$$1 \equiv x^2 = a^2 - b^2 \equiv -1 \pmod 4,$$

so b is even. We apply Theorem 7–1 again, this time to the equation

$$x^2 + b^2 = a^2,$$

and obtain

$$x = p^2 - q^2, \qquad b = 2pq, \qquad a = p^2 + q^2,$$

where $(p, q) = 1$, $p > q > 0$, and not both p and q are odd. From $y^2 = 2ab$ we have

$$y^2 = 4pq(p^2 + q^2).$$

Here any two of p, q and $p^2 + q^2$ are relatively prime, and hence each must be a square:

$$p = r^2, \qquad q = s^2, \qquad p^2 + q^2 = t^2,$$

whence $t > 1$ and

$$r^4 + s^4 = t^2$$

Now

$$x = r^4 - s^4, \qquad y = 2rst, \qquad z = a^2 + b^2 = r^8 + 6r^4s^4 + s^8,$$

so that

$$z > (r^4 + s^4)^2 = t^4,$$

or $t < z^{1/4}$. It follows that if one nonzero solution of $x^4 + y^4 = z^2$ were known, another solution r, s, t could be found for which $rst \neq 0$ and $1 < t < z^{1/4}$. If we started from r, s, t instead of x, y, z, a third solution r', s', t' could then be found such that $1 < t' < t^{1/4}$, and so on. But this would yield an infinite decreasing sequence of positive integers, $z, t, t', \ldots$, which is impossible. Thus there is no nonzero solution. ▲

7-4 The equation $x^2 - dy^2 = 1$. The Diophantine equation

$$x^2 - dy^2 = N$$

(where N and d are integers), commonly known as Pell's equation, was actually never considered by Pell; it was because of a mistake on Euler's part that Pell's name has been attached to it. The early Greek and Indian mathematicians had considered special cases, but Fermat was the first to treat it systematically. He said that he had shown, in the special case where $N = 1$ and $d > 0$ is not a perfect square, that there are infinitely many integral solutions x, y; as usual, he did not give a proof. The first published proof was given by Lagrange, who used the theory of continued fractions. Prior to this, Euler had shown that there are infinitely many solutions if there is one.

Regardless of the name given to it, the equation is of considerable importance in number theory. We saw at the end of the preceding chapter how it arises in connection with the units of real quadratic domains, a subject seemingly removed from Diophantine equations. It also plays a central role in the theory of indefinite binary quadratic forms, a more advanced branch of the theory of numbers. Even within the theory of Diophantine equations, Pell's equation is fundamental, because so many other equations can be reduced to it, or made to depend on it in some way. For example, knowledge of the solutions of Pell's equation is essential in finding integral solutions of the general quadratic equation

$$ax^2 + bxy + cy^2 + dx + ey + f = 0,$$

in which $a, b, \ldots, f$ are integers. For, writing the left side as a polynomial in x,

$$ax^2 + (by + d)x + cy^2 + ey + f = 0,$$

we see that if the equation is solvable for a certain y, the discriminant

$$(by + d)^2 - 4a(cy^2 + ey + f)$$

or, what is the same thing,

$$(b^2 - 4ac)y^2 + (2bd - 4ae)y + d^2 - 4af$$

must be a perfect square, say z^2. Setting

$$b^2 - 4ac = p, \qquad 2bd - 4ae = q, \qquad d^2 - 4af = r,$$

we have

$$py^2 + qy + r - z^2 = 0.$$

Again, the discriminant of this quadratic in y must be a perfect square, say

$$q^2 - 4p(r - z^2) = w^2.$$

Thus we are led to consider the Pell equation

$$w^2 - 4pz^2 = q^2 - 4pr;$$

once we know solutions of this equation, we can, at any rate, obtain rational solutions of the original quadratic equation.

It might also be mentioned that Pell's equation shares with the linear equation $ax + by = c$ a unique position among Diophantine equations in two unknowns. It was shown in 1929 by C. L. Siegel that these two equations, together with the equations derivable from them by certain transformations, are the only algebraic equations in two variables which can have infinitely many integral solutions!

Now to the solution. For the present we shall concern ourselves with the equation

$$x^2 - dy^2 = 1. \tag{1}$$

The case in which d is a negative integer is easily settled: if $d = -1$, then the only solutions are $\pm 1, 0$ and $0, \pm 1$, whereas if $d < -1$, the only solutions are $\pm 1, 0$. So from now on we may restrict our attention to equations of the form (1) with $d > 0$. If d is a square, then (1) can be written as

$$x^2 - (d'y)^2 = 1,$$

and since the only two squares which differ by 1 are 0 and 1, the only solutions in this case are $\pm 1, 0$. Suppose then that d is not a square.

Except for the trivial solutions $\pm 1, 0$, we have $xy \neq 0$ and hence four solutions, $\{x, y\}$, $\{x, -y\}$, $\{-x, y\}$, $\{-x, -y\}$, which are associated with one another in a simple way. Let us confine our attention for the moment to the positive solutions, in which $x > 0$ and $y > 0$. Equation (1) can be written in the form $(x - y\sqrt{d})(x + y\sqrt{d}) = 1$, or

$$x - y\sqrt{d} = \frac{1}{x + y\sqrt{d}}, \tag{2}$$

and for large x and y the right-hand side of this equation is very small. Hence $x - y\sqrt{d}$, or $y(x/y - \sqrt{d})$, is also small, which means that x/y is required to be a very good rational approximation to the irrational number $\sqrt{d}$. It must, in fact, be such a good approximation that even the product of the error $x/y - \sqrt{d}$ and the large number y is very small. This is a strong condition, and is satisfied only by exceptional fractions x/y. But it must be satisfied infinitely many times, if (1) is to have infinitely many solutions.

Conversely, if we could find positive integers x and y such that

$$0 < x - y\sqrt{d} < \frac{2}{x + y\sqrt{d}},$$

then we would have $0 < (x - y\sqrt{d})(x + y\sqrt{d}) = x^2 - dy^2 < 2$, and since $x^2 - dy^2$ is an integer, it would follow that $x^2 - dy^2 = 1$. Thus solutions of (1) give good approximations to $\sqrt{d}$, and sufficiently good approximations to d provide solutions of (1). As was seen in Chapter 5, the best approximations to an irrational number are furnished by the convergents of the continued fraction expansion of that number, and therefore we first look to see how (2) is related to the inequalities of Chapter 5.

THEOREM 7–3. If (1) holds, and x and y are positive, then x/y is a convergent of the continued fraction expansion of $\sqrt{d}$.

Proof: By (2), $x - y\sqrt{d} > 0$, so that $x/y > \sqrt{d}$. Hence (2) implies that

$$\left|\frac{x}{y} - \sqrt{d}\right| = \frac{1}{y^2\left(\frac{x}{y} + \sqrt{d}\right)} < \frac{1}{2y^2\sqrt{d}} < \frac{1}{2y^2},$$

and the result follows from Theorem 5–14. ▲

The converse of Theorem 7–3 is in general false. Instead, we have the following result, which will lead to solutions of (1) in a rather roundabout way.

THEOREM 7–4. Every convergent p_n/q_n of the continued fraction expansion of $\sqrt{d}$ provides a solution $x = p_n$, $y = q_n$ of one of the equations

$$x^2 - dy^2 = k,$$

where k ranges over the finitely many integers such that $|k| < 1 + 2\sqrt{d}$. In particular, one of these equations has infinitely many solutions.

Proof: By Theorem 5–12, we have

$$\left|\frac{p_n}{q_n} - \sqrt{d}\right| < \frac{1}{q_n^2},$$

and hence

$$|p_n - q_n\sqrt{d}| < \frac{1}{q_n}, \tag{3}$$

and

$$\frac{p_n}{q_n} < \sqrt{d} + \frac{1}{q_n^2} \leq \sqrt{d} + 1. \tag{4}$$

Thus, using first (3) and then (4), we obtain

$$|p_n^2 - dq_n^2| < \frac{p_n + q_n\sqrt{d}}{q_n} = \frac{p_n}{q_n} + \sqrt{d} \leq 2\sqrt{d} + 1. \blacktriangle$$

In the remainder of the discussion of Pell's equation, we shall have more occasion to speak of the combination $x + y\sqrt{d}$ than of the solution $\{x, y\}$. For this reason we pause for a moment to consider these irrational numbers as interesting objects in themselves. Let us designate by $Z[\sqrt{d}]$ the set of *all* numbers of the form $a + b\sqrt{d}$ in which a and b are integers, and use Greek letters to stand for the elements of $Z[\sqrt{d}]$. If $\alpha = a + b\sqrt{d}$, then a and b are called the *components* of α. For α and β in $Z[\sqrt{d}]$, the combinations $\alpha\beta$, $\alpha + \beta$, and $\alpha - \beta$ are again in $Z[\sqrt{d}]$, but β/α need not be. For if by the *conjugate* of $\alpha = a + b\sqrt{d}$ we mean the element $\bar{\alpha} = a - b\sqrt{d}$ of $Z[\sqrt{d}]$, then

$$\frac{\beta}{\alpha} = \frac{\beta\bar{\alpha}}{\alpha\bar{\alpha}} = \frac{\beta\bar{\alpha}}{a^2 - db^2},$$

and this is an element of $Z[\sqrt{d}]$ only if its components are integers, that is, only if $a^2 - db^2$ divides the components of $\beta\bar{\alpha}$.

If m is a positive integer, then we say that m *divides* α, or that $\alpha \equiv 0 \pmod{m}$, if m divides both components of α. We say that $\alpha \equiv \beta \pmod{m}$ if $\alpha - \beta \equiv 0 \pmod{m}$. By completely trivial modifications in the proofs given at the beginning of Chapter 3, we see that this kind of congruence is again an equivalence relation, that two congruences with the same modulus can be added or multiplied together, and that a congruence can be multiplied through by an arbitrary factor from $Z[\sqrt{d}]$. There is one slight change in that there are now m^2 residue classes, rather than m as before, because each of the two components can have any of m incongruent values.

Just as in the preceding chapter, we call the product $\alpha\bar{\alpha}$ the *norm* of α, and write $\mathbf{N}(\alpha)$, or simply $\mathbf{N}\alpha$. To ask for solutions of $x^2 - dy^2 = k$ is to ask for elements α of $Z[\sqrt{d}]$ such that $\mathbf{N}\alpha = k$. Clearly $\mathbf{N}\alpha\beta = (\alpha\beta)(\overline{\alpha\beta}) = \alpha\beta\bar{\alpha}\bar{\beta} = (\alpha\bar{\alpha})(\beta\bar{\beta}) = \mathbf{N}\alpha\mathbf{N}\beta$.

If the components of α are positive, then α, which is a real number as well as an element of $Z[\sqrt{d}]$, is larger than 1. The four elements of $Z[\sqrt{d}]$ which have the same components as α except for sign are α, $\bar{\alpha}$, $-\alpha$, and $-\bar{\alpha}$. If α has as components a positive solution of (1), then $\alpha\bar{\alpha} = 1$, and the four numbers just mentioned are α, $1/\alpha$, $-\alpha$, and $-1/\alpha$. Of these the first is larger than 1, the second is between 0 and 1, the third is smaller than -1, and the fourth lies between -1 and 0, so that the signs of x and y in (1) determine, and are determined by, the size of the associated α. To consider positive solutions of (1) is to consider elements $\alpha > 1$ of $Z[\sqrt{d}]$.

THEOREM 7–5. Equation (1) has at least one solution with $y \neq 0$.

Proof: According to Theorem 7–4, there is an integer k for which the equation $\mathbf{N}\alpha = k$ has infinitely many solutions α in $Z[\sqrt{d}]$. Since there are only finitely many residue classes (mod k) in $Z[\sqrt{d}]$, some residue class must contain at least two of these solutions (in fact, infinitely many!). Let us assume then that $\mathbf{N}\alpha_1 = \mathbf{N}\alpha_2 = k$ and $\alpha_1 \equiv \alpha_2 \pmod{k}$, but that $\alpha_1 \neq \alpha_2$. Then $\alpha_1\bar{\alpha}_2 \equiv \alpha_2\bar{\alpha}_2 \equiv 0 \pmod{k}$, so that $\beta = \alpha_1\bar{\alpha}_2/k$ is an element of $Z[\sqrt{d}]$; that is, it has integral components. Since

$$\mathbf{N}\beta = \beta\bar{\beta} = \frac{\alpha_1\bar{\alpha}_2 \cdot \bar{\alpha}_1\alpha_2}{k^2} = \frac{\mathbf{N}\alpha_1\mathbf{N}\alpha_2}{k^2} = 1,$$

β yields a solution of (1). If the second component of β were 0, then $\mathbf{N}\beta = 1$ would imply that $\beta = 1$, whence

$$\begin{aligned} \alpha_1\bar{\alpha}_2 &= k = \alpha_1\bar{\alpha}_1, \\ \bar{\alpha}_2 &= \bar{\alpha}_1, \\ \alpha_2 &= \alpha_1, \end{aligned}$$

contrary to hypothesis. ▲

THEOREM 7–6. If x_1, y_1 and x_2, y_2 are solutions of the Pell equation (1), then so are the integers x, y defined by the equation

$$(x_1 + y_1\sqrt{d})(x_2 + y_2\sqrt{d}) = x + y\sqrt{d}. \tag{5}$$

Proof: The theorem merely asserts that if $\mathbf{N}\alpha = 1$ and $\mathbf{N}\beta = 1$, then $\mathbf{N}(\alpha\beta) = 1$. ▲

Theorem 7–5 shows that there is an α in $Z[\sqrt{d}]$ such that $\alpha > 1$ and $\mathbf{N}\alpha = 1$, and Theorem 7–6 demonstrates that all the powers α^n give solutions of (1). Since $\alpha < \alpha^2 < \alpha^3 < \dots$, we see that (1) has infinitely many distinct solutions. The next theorem shows that all the solutions arise, in essence, from a single one.

THEOREM 7–7. If x_1, y_1 is the minimal positive solution of equation (1), then every solution x, y is given by the equation

$$x + y\sqrt{d} = \pm(x_1 + y_1\sqrt{d})^n, \tag{6}$$

where n can assume any integral value, positive, negative, or zero.

Because of this theorem, the minimal positive solution of (1) is sometimes called the *fundamental* solution.

Proof: We have already seen that the four numbers α^n, $1/\alpha^n$, $-\alpha^n$, and $-1/\alpha^n$ give four solutions differing only in the signs of x and y, so we need only show that every $\alpha > 1$ such that $\mathbf{N}\alpha = 1$ is of the form $\alpha = \delta^n$ for suitable positive integer n. Here δ is the fundamental solution, and therefore it is the smallest element of $Z[\sqrt{d}]$ which is larger than 1 and has norm 1.

Since $\alpha > 1$ and δ is minimal, we have $\alpha \geq \delta$. Hence there is a positive integer n such that $\delta^n \leq \alpha < \delta^{n+1}$. Now $\alpha/\delta^n = \alpha\bar{\delta}^n$ is in $Z[\sqrt{d}]$, and $\mathbf{N}(\alpha/\delta^n) = 1$. In other words, the number $\alpha/\delta^n = \beta$ gives an integral solution of (1). From the definition of n it follows that $1 \leq \beta < \delta$, and by the definition of δ we cannot have $1 < \beta < \delta$. Hence $\beta = 1$, and $\alpha = \delta^n$. ▲

PROBLEMS

1. Modify the proof of Theorem 7–3 to show that if $x^2 - dy^2 = N$, $0 < N < \sqrt{d}$, then x/y is a convergent of the continued fraction expansion of $\sqrt{d}$.

2. Show that if $x^2 - dy^2 = -N$, $0 < N < \sqrt{d}$, then x/y is a convergent of the continued fraction expansion of $\sqrt{d}$. [*Hint:* Show that

$$0 < y - \frac{x}{\sqrt{d}} < \frac{1}{x\left(1 + \frac{y\sqrt{d}}{x}\right)},$$

and deduce that

$$\left|\frac{1}{\sqrt{d}} - \frac{y}{x}\right| < \frac{1}{2x^2}.$$

Then use the simple relation which exists between the convergents of $\sqrt{d}$ and those of $1/\sqrt{d}$, as indicated in Problem 7, Section 5-4.]

7-5 The equation $x^2 - dy^2 = -1$. This equation differs from $x^2 - dy^2 = 1$ (+1 on the right hand side) in that it may well have no solutions at all with $y \neq 0$. For if $x^2 - dy^2 = -1$, then $x^2 \equiv -1 \pmod{d}$, so that -1 must be a quadratic residue of d. This is certainly not the case, for example, if d is any prime $p \equiv 3 \pmod 4$, according to Theorem 3-15. On the other hand, if the equation is solvable, the structure of the set of solutions is similar to that considered in the preceding section.

THEOREM 7-8. Let d be a positive nonsquare integer. Then if the equation

$$z^2 - dt^2 = -1 \tag{7}$$

is solvable, and if $\gamma = z_1 + t_1\sqrt{d}$ is the minimal positive solution, a general solution is given by

$$z + t\sqrt{d} = \pm\gamma^{2n+1}, \qquad n = 0, \pm 1, \pm 2, \ldots.$$

With the earlier notation, $\delta = \gamma^2$.

Proof: We prove the second assertion first. It is clear that γ^2 is a solution of (1), since $\mathbf{N}(\gamma^2) = (\mathbf{N}\gamma)^2 = 1$, and hence $1 < \delta < \gamma^2$, by the definition of δ. Since $1/\gamma = -\bar{\gamma}$, we have

$$\frac{1}{\gamma} < -\delta\bar{\gamma} \leq \gamma, \tag{8}$$

and $\mathbf{N}(-\delta\bar{\gamma}) = \mathbf{N}\delta\mathbf{N}\bar{\gamma} = -1$. Thus $-\delta\bar{\gamma} = \beta$ is a solution of (7), and in particular $\beta \neq 1$. Taking reciprocals in (8) yields $1/\gamma \leq 1/\beta < \gamma$, and hence either

$$1 < \beta \leq \gamma \qquad \text{or} \qquad 1 < \frac{1}{\beta} < \gamma.$$

Using the minimality of γ, we conclude that $\beta = \gamma$, and hence that $\delta = \gamma^2$.

Now suppose that β is any solution of (7); we can again restrict attention to the case $\beta > 1$. Then as in the proof of Theorem 7-7, there is a positive integer n such that

$$1 \leq \beta\,\delta^{-n} < \delta = \gamma^2,$$

and dividing through by γ, we obtain

$$\gamma^{-1} \leq \alpha < \gamma,$$

where $\alpha = \beta\,\delta^{-n}\gamma^{-1}$ is a solution of (1). Since $1 < \gamma < \delta$, the last inequality implies that $\delta^{-1} < \alpha < \delta$, so that $\alpha = 1$ and $\beta = \delta^n\gamma = \gamma^{2n+1}$. ▲

Problems

1. Let ξ_n be the "fractional part" of $n\sqrt{2}$:

$$\xi_n = n\sqrt{2} - [n\sqrt{2}],$$

and let t be a positive real number independent of n. (a) Show that if $t < 1/2\sqrt{2}$, then $n\xi_n > t$ for all sufficiently large t. (b) Show that if $t > 1/2\sqrt{2}$, then for a suitable sequence $n_1 \leq n_2 < \cdots$ of positive integers, $n_k\xi_{n_k} < t$.

7–6 Pell's equation and continued fractions. We can now make more precise the connection established in Section 7–4 between the convergents to $\sqrt{d}$ and the solutions of $x^2 - dy^2 = \pm 1$. By Theorem 7–3 and Problem 2 of Section 7–4, we know that all such solutions are to be found among the convergents of $\sqrt{d}$. The problem is to determine how far out one must hunt to find the minimal solutions. With respect to the equation with $+1$ on the right, the situation is not too bad; we know that a solution exists, and that it must show up eventually among the quantities $p_n^2 - dq_n^2$. But for the equation with -1 on the right, we do not yet have a criterion for deciding whether the equation is solvable, and it would be desirable to know that if no solution turns up among the first M convergents, for suitable M depending on d, then there is no solution. The following theorem clarifies the situation.

Theorem 7–9. The sequence $\{p_n^2 - dq_n^2\}$ is eventually periodic, the periodicity beginning (at the latest) at the index preceding that at which the sequence of partial quotients becomes periodic. The length of the period of the first sequence is at most twice that of the second.

Proof: With $\xi = x = \sqrt{d}$, equation (11) of Chapter 5 yields

$$\sqrt{d} = \frac{p_{k-1}\xi_k + p_{k-2}}{q_{k-1}\xi_k + q_{k-2}}. \tag{9}$$

Solving for ξ_k and rationalizing the denominator, we can write

$$\xi_k = \frac{\sqrt{d} + r_k}{s_k},$$

where r_k and s_k are rational numbers. Substituting this into (9), and then replacing k by $k + 1$ throughout, we have

$$\sqrt{d} = \frac{p_k(\sqrt{d} + r_{k+1}) + p_{k-1}s_{k+1}}{q_k(\sqrt{d} + r_{k+1}) + q_{k-1}s_{k+1}},$$

or

$$(q_k r_{k+1} + q_{k-1}s_{k+1} - p_k)\sqrt{d} - (p_{k-1}s_{k+1} + p_k r_{k+1} - q_k d) = 0.$$

The rational and irrational parts must separately be zero, so

$$q_k r_{k+1} + q_{k-1}s_{k+1} = p_k, \qquad p_k r_{k+1} + p_{k-1}s_{k+1} = q_k d.$$

The determinant of this system is $q_k p_{k-1} - q_{k-1}p_k = (-1)^k$, and hence

$$\begin{aligned} r_{k+1} &= (-1)^k(p_k p_{k-1} - q_k q_{k-1} d), \\ s_{k+1} &= (-1)^k(q_k^2 d - p_k^2). \end{aligned} \tag{10}$$

Now the numbers r_k and s_k are uniquely determined by ξ_k; since $\{\xi_k\}$ is eventually periodic, the same is true of $\{s_k\}$, and the eventual periodicity of $\{p_k^2 - dq_k^2\}$ follows from (10). In fact, $\{s_k\}$ becomes periodic at the same index as $\{\xi_k\}$, so $\{s_{k+1}\}$ becomes periodic at the preceding index. Finally, the length of the period of $\{(-1)^{k-1}s_k\}$ is clearly at most twice that of $\{s_k\}$. ▲

As an example, consider the equations $x^2 - 7y^2 = \pm 1$. Continuing the computations of Section 5–5, we can add a new row to the table occurring there:

k:	0	1	2	3	4	5	6 . . . ,
$p_k^2 - 7q_k^2$:	−3	2	−3	1	−3	2	−3 . . . .

In this case the sequence $\{p_k^2 - 7q_k^2\}$ is periodic from the beginning, with period length 4. (The sequence of partial quotients is periodic after a_0, with period length 4.) The minimal positive solution of $x^2 - 7y^2 = 1$ is $x = 8$, $y = 3$. There is no solution of $x^2 - 7y^2 = -1$, since none shows up in the first period.

Consider instead $x^2 - 5y^2 = \pm 1$. The continued fraction expansion of $\sqrt{5}$ is $\{2; 4, 4, 4, \ldots\}$, the convergents are 2/1, 9/4, 38/17, . . . , and the sequence $\{p_k^2 - 5q_k^2\}$ is $\{-1, 1, -1, 1, \ldots\}$. The period length of the latter sequence is twice that of $\{a_k\}$, and both Pell equations are solvable.

We shall not prove it, but what has happened in these examples is typical: the continued fraction expansion of $\sqrt{d}$ is always periodic after a_0, so that $\{p_k^2 - dq_k^2\}$ is always periodic from the beginning, and the equation $x^2 - dy^2 = -1$ is solvable if and only if the period length of $\{a_k\}$ is odd.

Problems

1. Find the minimum positive solutions of
   (a) $x^2 - 94y^2 = 1$, (b) $x^2 - 95y^2 = 1$.

2. Using the results of Problems 1 and 2 of Section 7–4, find all the N with $|N| < 10$ for which the equation $x^2 - 95y^2 = N$ is solvable.

APPENDIX

Table of Primes

2	3	5	7	11	13	17	19	23	29	31	37	41
43	47	53	59	61	67	71	73	79	83	89	97	101
103	107	109	113	127	131	137	139	149	151	157	163	167
173	179	181	191	193	197	199	211	223	227	229	233	239
241	251	257	263	269	271	277	281	283	293	307	311	313
317	331	337	347	349	353	359	367	373	379	383	389	397
401	409	419	421	431	433	439	443	449	457	461	463	467
479	487	491	499	503	509	521	523	541	547	557	563	569
571	577	587	593	599	601	607	613	617	619	631	641	643
647	653	659	661	673	677	683	691	701	709	719	727	733
739	743	751	757	761	769	773	787	797	809	811	821	823
827	829	839	853	857	859	863	877	881	883	887	907	911
919	929	937	941	947	953	967	971	977	983	991	997	1009

PRIMITIVE ROOTS

p	Smallest primitive root g of p	p	g
3	2	43	3
5	2	47	5
7	3	53	2
11	2	59	2
13	2	61	2
17	3	67	2
19	2	71	7
23	5	73	5
29	2	79	3
31	3	83	2
37	2	89	3
41	6	97	5

Greek Alphabet

A	α	Alpha	N	ν	Nu
B	β	Beta	Ξ	ξ	Xi
Γ	γ	Gamma	O	o	Omicron
Δ	δ	Delta	Π	π	Pi
E	ϵ	Epsilon	P	ρ	Rho
Z	ζ	Zeta	Σ	σ	Sigma
H	η	Eta	T	τ	Tau
Θ	θ	Theta	Υ	υ	Upsilon
I	ι	Iota	Φ	ϕ, φ	Phi
K	κ	Kappa	X	χ	Chi
Λ	λ	Lambda	Ψ	ψ	Psi
M	μ	Mu	Ω	ω	Omega

Supplementary Reading

Davenport, H., *The Higher Arithmetic*, London: Hutchinson and Co. (Publishers) Ltd., 1952. Reprinted, Dover Publications, Inc., New York, 1983.

Griffin, H., *Elementary Theory of Numbers*, New York: McGraw-Hill Book Co., Inc., 1954.

Hardy, G. H. and E. M. Wright, *An Introduction to the Theory of Numbers*, 4th edition, New York: Oxford University Press, 1960.

Jones, B. W., *The Theory of Numbers*, New York: Rinehart and Co., Inc., 1955.

Nagell, T., *Introduction to the Theory of Numbers*, New York: John Wiley and Sons, Inc., 1951.

Niven, I. and H. S. Zuckerman, *An Introduction to the Theory of Numbers*, New York: John Wiley and Sons, Inc., 1960.

Ore, Ø., *Number Theory and its History*, New York: McGraw-Hill Book Company, Inc., 1948. Reprinted, Dover Publications, Inc., New York, 1987.

Stewart, B. M., *Theory of Numbers*, New York: The Macmillan Company, 1952.

Wright, H. N., *Theory of Numbers*, New York: John Wiley and Sons, Inc., 1939.

ANSWERS TO SELECTED PROBLEMS

ANSWERS TO SELECTED PROBLEMS

Section 1–3

7. $\beta = (1 + \sqrt{5})/2, \quad c = 1/\beta^2$ 8. $\beta = 3, \quad c = 2/3$

Section 1–5

2. $(204734\beta 8)_{12}$

Section 2–2

6. (a) 35; $x = -83, y = 32$ (b) 1; $x = -1013, y = 534$
8. (a) 17 (c) 19

Section 2–4

1. $x = -3 + 7t, \quad y = 5 - 8t$
2. There are five solutions, for which $x = 11, 31, 51, 71, 91$.
5. A general solution is

$$\begin{aligned} x &= 45 - 267r + 166s, \\ y &= 3r + 4s, \\ z &= -37 + 211r - 159s, \end{aligned}$$

where r and s are arbitrary integers.
7. 32.65

Section 3–5

1. (a) $x \equiv 7, 42$, or $77 \pmod{105}$ (c) $x \equiv 2, 9, 16, \ldots, 100 \pmod{105}$
2. The solutions (mod 18) are:

1, 2	3, 3	5, 4	7, 5	9, 0	11, 1	13, 2	15, 3	17, 4
1, 8	3, 9	5, 10	7, 11	9, 6	11, 7	13, 8	15, 9	17, 10
1, 14	3, 15	5, 16	7, 17	9, 12	11, 13	13, 14	15, 15	17, 16

3. $x \equiv 173 \pmod{210}$

5 (a). The system $x_1 \equiv \{1, 0, 0\} \pmod{\{27, 25, 8\}}$ is equivalent to $x_1 \equiv \{1, 0\}$ $\pmod{\{27, 200\}}$, or setting $x_1 = 200y_1$, we find that it is equivalent to $200y_1 \equiv 1$ $\pmod{27}$. This has the solution $y_1 \equiv 5 \pmod{27}$, whence $x_1 \equiv 1000 \pmod{5400}$. Similarly, we write

$$\begin{aligned} x_2 &\equiv \{0, 1, 0\} \pmod{\{27, 25, 8\}}, \\ x_2 &\equiv \{1, 0\} \pmod{\{25, 216\}}, \\ x_2 &= 216y_2, \\ y_2 &\equiv 11 \pmod{25}, \\ x_2 &\equiv 2376 \pmod{5400}, \end{aligned}$$

and in the same way we obtain $x_2 \equiv 2025 \pmod{5400}$. For this system, this yields $x \equiv 10x_1 + 2x_2 + 3x_3 \equiv 4627 \pmod{5400}$.

6. (a) 529 (b) 397 (c) 100

Section 3–7

2. $-1, 1, -1$
5. (a) No solution (b) $x \equiv 10$ or $12 \pmod{13}$

Section 4–1

6. (a) 3 (b) 30 (c) 4

Section 4–2

6. 2, 3, 8, 12, 13, 17, 22, 23
7. 5, 7, 10, 11, 14, 15, 17, 19, 20, 21

Section 4–3

1. (a) $x \equiv 10 \pmod{29}$ (c) $x \equiv 2, 27 \pmod{29}$ (e) $x \equiv 9, 24 \pmod{29}$ (h) $x \equiv 8, 10, 12, 15, 18, 26, 27 \pmod{29}$
2. (a) No (b) Yes

Section 5–1

1. (a) $\{2; 3, 5, 2\}$ (c) $\{5;\}$ 2. (a) $\pi = \{3; 7, 15, 1, \pi_4\}$, where

$$\pi_4 = \frac{106\pi - 333}{355 - 113\pi}$$

The approximation $\pi \approx 3.14159265$ gives $\pi_4 \approx 285.4$, whereas a better approximation to π shows that $[\pi_4] = 292$. (b) $\{1; 1, 2, 1, 2/(\sqrt{3} - 1)\}$
7. 2, 24

Section 5–2

1. (a) $\frac{301}{96}$ (c) $\frac{8}{5}$, and more generally u_{n+1}/u_n, where the u_k are the Fibonacci numbers.

Section 5–3

1. (a) $x = 276 + 2971t$, $y = 293 + 3154t$

Section 6–3

1. (a) $-1 + 5i$ (c) 1 [Note that if $(\alpha, \beta) = \delta$, then $\mathbf{N}\delta | (\mathbf{N}\alpha, \mathbf{N}\beta)$.]
2. $a \equiv b \pmod{2}$
3. (a) $-1 + 5i = 1 \cdot (33 + 17i) - (2 + i)(16 - 2i)$

Section 6–4

2. (a) The first-quadrant integers α with $\mathbf{N}\alpha \leq 9$ are i, $2i$, $3i$, $1 + i$, $1 + 2i$, $2 + i$, $2 + 2i$. Of these, $2i$, $1 + i$, and $2 + 2i$ are multiples of $1 + i$. The remaining numbers must be primes or units, so the Gaussian primes with norms ≤ 9 are exactly the associates of $3i$, $1 + 2i$, and $2 + i$.
3. (a) Prime (b) $(1 + i)(2 - 5i)$

Section 6–5

1. The Gaussian primes with norms ≤ 50 are the associates of $1 + i$, 3, 7, $1 \pm 2i$, $2 \pm 3i$, $4 \pm i$, $5 \pm 2i$, $6 \pm i$, $5 \pm 4i$.
2. (a) $(1 - 2i)(-1 + 4i)$ (b) $(1 + i)(6 - i)$

Section 7–6

1. (a) $x = 2{,}143{,}295$, $y = 221{,}064$ (b) $x = 39$, $y = 4$
2. $N = 1, 5$

INDEX

INDEX

A CATALOG OF SELECTED
DOVER BOOKS
IN SCIENCE AND MATHEMATICS

A CATALOG OF SELECTED

DOVER BOOKS

IN SCIENCE AND MATHEMATICS

QUALITATIVE THEORY OF DIFFERENTIAL EQUATIONS, V.V. Nemytskii and V.V. Stepanov. Classic graduate-level text by two prominent Soviet mathematicians covers classical differential equations as well as topological dynamics and ergodic theory. Bibliographies. 523pp. 5⅜ × 8½. 65954-2 Pa. $10.95

MATRICES AND LINEAR ALGEBRA, Hans Schneider and George Phillip Barker. Basic textbook covers theory of matrices and its applications to systems of linear equations and related topics such as determinants, eigenvalues and differential equations. Numerous exercises. 432pp. 5⅜ × 8½. 66014-1 Pa. $10.95

QUANTUM THEORY, David Bohm. This advanced undergraduate-level text presents the quantum theory in terms of qualitative and imaginative concepts, followed by specific applications worked out in mathematical detail. Preface. Index. 655pp. 5⅜ × 8½. 65969-0 Pa. $13.95

ATOMIC PHYSICS (8th edition), Max Born. Nobel laureate's lucid treatment of kinetic theory of gases, elementary particles, nuclear atom, wave-corpuscles, atomic structure and spectral lines, much more. Over 40 appendices, bibliography. 495pp. 5⅜ × 8½. 65984-4 Pa. $12.95

ELECTRONIC STRUCTURE AND THE PROPERTIES OF SOLIDS: The Physics of the Chemical Bond, Walter A. Harrison. Innovative text offers basic understanding of the electronic structure of covalent and ionic solids, simple metals, transition metals and their compounds. Problems. 1980 edition. 582pp. 6⅛ × 9¼. 66021-4 Pa. $15.95

BOUNDARY VALUE PROBLEMS OF HEAT CONDUCTION, M. Necati Özisik. Systematic, comprehensive treatment of modern mathematical methods of solving problems in heat conduction and diffusion. Numerous examples and problems. Selected references. Appendices. 505pp. 5⅜ × 8½. 65990-9 Pa. $12.95

A SHORT HISTORY OF CHEMISTRY (3rd edition), J.R. Partington. Classic exposition explores origins of chemistry, alchemy, early medical chemistry, nature of atmosphere, theory of valency, laws and structure of atomic theory, much more. 428pp. 5⅜ × 8½. (Available in U.S. only) 65977-1 Pa. $10.95

A HISTORY OF ASTRONOMY, A. Pannekoek. Well-balanced, carefully reasoned study covers such topics as Ptolemaic theory, work of Copernicus, Kepler, Newton, Eddington's work on stars, much more. Illustrated. References. 521pp. 5⅜ × 8½. 65994-1 Pa. $12.95

PRINCIPLES OF METEOROLOGICAL ANALYSIS, Walter J. Saucier. Highly respected, abundantly illustrated classic reviews atmospheric variables, hydrostatics, static stability, various analyses (scalar, cross-section, isobaric, isentropic, more). For intermediate meteorology students. 454pp. 6⅛ × 9¼. 65979-8 Pa. $14.95

CHALLENGING MATHEMATICAL PROBLEMS WITH ELEMENTARY SOLUTIONS, A.M. Yaglom and I.M. Yaglom. Over 170 challenging problems on probability theory, combinatorial analysis, points and lines, topology, convex polygons, many other topics. Solutions. Total of 445pp. 5⅜ × 8½. Two-vol. set.
Vol. I 65536-9 Pa. $7.95
Vol. II 65537-7 Pa. $6.95

FIFTY CHALLENGING PROBLEMS IN PROBABILITY WITH SOLUTIONS, Frederick Mosteller. Remarkable puzzlers, graded in difficulty, illustrate elementary and advanced aspects of probability. Detailed solutions. 88pp. 5⅜ × 8½.
65355-2 Pa. $4.95

EXPERIMENTS IN TOPOLOGY, Stephen Barr. Classic, lively explanation of one of the byways of mathematics. Klein bottles, Moebius strips, projective planes, map coloring, problem of the Koenigsberg bridges, much more, described with clarity and wit. 43 figures. 210pp. 5⅜ × 8½. 25933-1 Pa. $5.95

RELATIVITY IN ILLUSTRATIONS, Jacob T. Schwartz. Clear nontechnical treatment makes relativity more accessible than ever before. Over 60 drawings illustrate concepts more clearly than text alone. Only high school geometry needed. Bibliography. 128pp. 6⅛ × 9¼. 25965-X Pa. $6.95

AN INTRODUCTION TO ORDINARY DIFFERENTIAL EQUATIONS, Earl A. Coddington. A thorough and systematic first course in elementary differential equations for undergraduates in mathematics and science, with many exercises and problems (with answers). Index. 304pp. 5⅜ × 8½. 65942-9 Pa. $8.95

FOURIER SERIES AND ORTHOGONAL FUNCTIONS, Harry F. Davis. An incisive text combining theory and practical example to introduce Fourier series, orthogonal functions and applications of the Fourier method to boundary-value problems. 570 exercises. Answers and notes. 416pp. 5⅜ × 8½. 65973-9 Pa. $9.95

THE THEORY OF BRANCHING PROCESSES, Theodore E. Harris. First systematic, comprehensive treatment of branching (i.e. multiplicative) processes and their applications. Galton-Watson model, Markov branching processes, electron-photon cascade, many other topics. Rigorous proofs. Bibliography. 240pp. 5⅜ × 8½. 65952-6 Pa. $6.95

AN INTRODUCTION TO ALGEBRAIC STRUCTURES, Joseph Landin. Superb self-contained text covers "abstract algebra": sets and numbers, theory of groups, theory of rings, much more. Numerous well-chosen examples, exercises. 247pp. 5⅜ × 8½. 65940-2 Pa. $7.95

Prices subject to change without notice.

Available at your book dealer or write for free Mathematics and Science Catalog to Dept. GI, Dover Publications, Inc., 31 East 2nd St., Mineola, N.Y. 11501. Dover publishes more than 175 books each year on science, elementary and advanced mathematics, biology, music, art, literature, history, social sciences and other areas.